BIOLOGICAL AND MEDICAL PHYSICS,
BIOMEDICAL ENGINEERING

BIOLOGICAL AND MEDICAL PHYSICS, BIOMEDICAL ENGINEERING

The fields of biological and medical physics and biomedical engineering are broad, multidisciplinary and dynamic. They lie at the crossroads of frontier research in physics, biology, chemistry, and medicine. The Biological and Medical Physics, Biomedical Engineering Series is intended to be comprehensive, covering a broad range of topics important to the study of the physical, chemical and biological sciences. Its goal is to provide scientists and engineers with textbooks, monographs, and reference works to address the growing need for information.

Books in the series emphasize established and emergent areas of science including molecular, membrane, and mathematical biophysics; photosynthetic energy harvesting and conversion; information processing; physical principles of genetics; sensory communications; automata networks, neural networks, and cellular automata. Equally important will be coverage of applied aspects of biological and medical physics and biomedical engineering such as molecular electronic components and devices, biosensors, medicine, imaging, physical principles of renewable energy production, advanced prostheses, and environmental control and engineering.

Editor-in-Chief:

Elias Greenbaum, Oak Ridge National Laboratory, Oak Ridge, Tennessee, USA

Editorial Board:

Masuo Aizawa, Department of Bioengineering, Tokyo Institute of Technology, Yokohama, Japan

Olaf S. Andersen, Department of Physiology, Biophysics & Molecular Medicine, Cornell University, New York, USA

Robert H. Austin, Department of Physics, Princeton University, Princeton, New Jersey, USA

James Barber, Department of Biochemistry, Imperial College of Science, Technology and Medicine, London, England

Howard C. Berg, Department of Molecular and Cellular Biology, Harvard University, Cambridge, Massachusetts, USA

Victor Bloomfield, Department of Biochemistry, University of Minnesota, St. Paul, Minnesota, USA

Robert Callender, Department of Biochemistry, Albert Einstein College of Medicine, Bronx, New York, USA

Britton Chance, Department of Biochemistry/ Biophysics, University of Pennsylvania, Philadelphia, Pennsylvania, USA

Steven Chu, Department of Physics, Stanford University, Stanford, California, USA

Louis J. DeFelice, Department of Pharmacology, Vanderbilt University, Nashville, Tennessee, USA

Johann Deisenhofer, Howard Hughes Medical Institute, The University of Texas, Dallas, Texas, USA

George Feher, Department of Physics, University of California, San Diego, La Jolla, California, USA

Hans Frauenfelder, CNLS, MS B258, Los Alamos National Laboratory, Los Alamos, New Mexico, USA

Ivar Giaever, Rensselaer Polytechnic Institute, Troy, New York, USA

Sol M. Gruner, Department of Physics, Princeton University, Princeton, New Jersey, USA

Judith Herzfeld, Department of Chemistry, Brandeis University, Waltham, Massachusetts, USA

Pierre Joliot, Institute de Biologie Physico-Chimique, Fondation Edmond de Rothschild, Paris, France

Lajos Keszthelyi, Institute of Biophysics, Hungarian Academy of Sciences, Szeged, Hungary

Robert S. Knox, Department of Physics and Astronomy, University of Rochester, Rochester, New York, USA

Aaron Lewis, Department of Applied Physics, Hebrew University, Jerusalem, Israel

Stuart M. Lindsay, Department of Physics and Astronomy, Arizona State University, Tempe, Arizona, USA

David Mauzerall, Rockefeller University, New York, New York, USA

Eugenie V. Mielczarek, Department of Physics and Astronomy, George Mason University, Fairfax, Virginia, USA

Markolf Niemz, Klinikum Mannheim, Mannheim, Germany

V. Adrian Parsegian, Physical Science Laboratory, National Institutes of Health, Bethesda, Maryland, USA

Linda S. Powers, NCDMF: Electrical Engineering, Utah State University, Logan, Utah, USA

Earl W. Prohofsky, Department of Physics, Purdue University, West Lafayette, Indiana, USA

Andrew Rubin, Department of Biophysics, Moscow State University, Moscow, Russia

Michael Seibert, National Renewable Energy Laboratory, Golden, Colorado, USA

David Thomas, Department of Biochemistry, University of Minnesota Medical School, Minneapolis, Minnesota, USA

Samuel J. Williamson, Department of Physics, New York University, New York, New York, USA

Y. Takeuchi Y. Iwasa K. Sato (Eds.)

Mathematics for Ecology and Environmental Sciences

With 26 Figures

 Springer

Prof. Yasuhiro Takeuchi
Shizuoka University
Faculty of Engineering
Department of Systems Engineering
Hamamatsu 3-5-1
432-8561 Shizuoka
Japan
email: takeuchi@sys.eng.shizuoka.ac.jp

Prof. Yoh Iwasa
Kyushu University
Department of Biology
812-8581 Fukuoka
Japan
e-mail: yiwasscb@mbox.nc.kyushu-u.ac.jp

Dr. Kazunori Sato
Shizuoka University
Faculty of Engineering
Department of Systems Engineering
Hamamatsu 3-5-1
432-8561 Shizuoka
Japan
email: sato@sys.eng.shizuoka.ac.jp

Library of Congress Cataloging in Publication Data: 2006931399

ISSN 1618-7210

ISBN-10 3-540-34427-6 Springer Berlin Heidelberg New York

ISBN-13 978-3-540-34427-8 Springer Berlin Heidelberg New York

This work is subject to copyright. All rights are reserved, whether the whole or part of the material is concerned, specifically the rights of translation, reprinting, reuse of illustrations, recitation, broadcasting, reproduction on microfilm or in any other way, and storage in data banks. Duplication of this publication or parts thereof is permitted only under the provisions of the German Copyright Law of September 9, 1965, in its current version, and permission for use must always be obtained from Springer. Violations are liable to prosecution under the German Copyright Law.

Springer is a part of Springer Science+Business Media

springer.com

© Springer-Verlag Berlin Heidelberg 2007

The use of general descriptive names, registered names, trademarks, etc. in this publication does not imply, even in the absence of a specific statement, that such names are exempt from the relevant protective laws and regulations and therefore free for general use.

Cover concept by eStudio Calamar Steinen

Cover production: WMXDesign GmbH, Heidelberg
Production: LE-TEX Jelonek, Schmidt, Vöckler GbR, Leipzig

Printed on acid-free paper   SPIN 10995792   57/3141/NN - 5 4 3 2 1 0

Preface

Dynamical systems theory in mathematical biology and environmental science has attracted much attention from many scientific fields as well as mathematics. For example, "chaos" is one of its typical topics. Recently the preservation of endangered species has become one of the most important issues in biology and environmental science, because of the recent rapid loss of biodiversity in the world. In this respect, permanence or persistence, new concepts in dynamical systems theory, seem important. These concepts give a new aspect in mathematics that includes various nonlinear phenomena such as chaos and phase transition, as well as the traditional concepts of stability and oscillation. Permanence and persistence analyses are expected not only to develop as new fields in mathematics but also to provide useful measures of robust survival for biological species in conservation biology and ecosystem management. Thus the study of dynamical systems will hopefully lead us to a useful policy for bio-diversity problems and the conservation of endangered species. The above fact brings us to recognize the importance of collaborations among mathematicians, biologists, environmental scientists and many related scientists as well. Mathematicians should establish a mathematical basis describing the various problems that appear in the dynamical systems of biology, and feed back their work to biology and environmental sciences. Biologists and environmental scientists should clarify/build the model systems that are important in their own global biological and environmental problems. In the end mathematics, biology and environmental sciences develop together.

The International Symposium "Dynamical Systems Theory and Its Applications to Biology and Environmental Sciences", held at Hamamatsu, Japan, March 14th–17th, 2004, under the chairmanship of one of the editors (Y.T.), gave the editors the idea for the book *Mathematics for Ecology and Environmental Sciences* and the chapters include material presented at the symposium as the invited lectures.

The editors asked authors of each chapter to follow some guidelines:

1. to keep in mind that each chapter will be read by many non-experts, who do not have background knowledge of the field;
2. at the beginning of each chapter, to explain the biological background of the modeling and theoretical work. This need not include detailed information about the biology, but enough knowledge to understand the model in question;
3. to review and summarize the previous theoretical and mathematical works and explain the context in which their own work is placed;
4. to explain the meaning of each term in the mathematical models, and the reason why the particular functional form is chosen, what is different from other authors' choices etc. What is obvious for the author may not be obvious for general readers;
5. then to present the mathematical analysis, which can be the main part of each chapter. If it is too technical, only the results and the main points of the technique of the mathematical analysis should be presented, rather than of showing all the steps of mathematical proof;
6. in the end of each chapter, to have a section ("Discussion") in which the author discusses biological implications of the outcome of the mathematical analysis (in addition to mathematical discussion).

Mathematics for Ecology and Environmental Sciences includes a wide variety of stimulating topics in mathematical and theoretical modeling and techniques to analyze the models in ecology and environmental sciences. It is hoped that the book will be useful as a source of future research projects on aspects of mathematical or theoretical modeling in ecology and environmental sciences. It is also hoped that the book will be useful to graduate students in the mathematical and biological sciences as well as to those in some areas of engineering and medicine. Readers should have had a course in calculus, and a knowledge of basic differential equations would be helpful.

We are especially pleased to acknowledge with gratitude the sponsorship and cooperation of Ministry of Education, Sports, Science and Technology, The Japanese Society for Mathematical Biology, The Society of Population Ecology, Mathematical Society of Japan, Japan Society for Industrial and Applied Mathematics, The Society for the Study of Species Biology, The Ecological Society of Japan, Society of Evolutionary Studies, Japan, Hamamatsu City and Shizuoka University, jointly with its Faculty of Engineering; Department of Systems Engineering.

Special thanks should also go to Keita Ashizawa for expert assistance with TEX. Drs. Claus Ascheron and Angela Lahee, the editorial staff of Springer-Verlag in Heidelberg, are warmly thanked.

Shizouka, *Yasuhiro Takeuchi*
Fukuoka, *Yoh Iwasa*
June 2006 *Kazunori Sato*

Contents

1 Ecology as a Modern Science
Kazunori Sato, Yoh Iwasa, Yasuhiro Takeuchi 1

**2 Physiologically Structured Population Models:
Towards a General Mathematical Theory**
Odo Diekmann, Mats Gyllenberg, Johan Metz 5

3 A Survey of Indirect Reciprocity
Hannelore Brandt, Hisashi Ohtsuki, Yoh Iwasa, Karl Sigmund 21

4 The Effects of Migration on Persistence and Extinction
Jingan Cui, Yasuhiro Takeuchi 51

**5 Sexual Reproduction Process
on One-Dimensional Stochastic Lattice Model**
Kazunori Sato 81

6 A Mathematical Model of Gene Transfer in a Biofilm
Mudassar Imran, Hal L. Smith 93

7 Nonlinearity and Stochasticity in Population Dynamics
J. M. Cushing 125

8 The Adaptive Dynamics of Community Structure
*Ulf Dieckmann, Åke Brännström,
Reinier HilleRisLambers, Hiroshi C. Ito* 145

Index 179

List of Contributors

Hannelore Brandt
Fakultät für Mathematik,
Nordbergstrasse 15, 1090 Wien,
Austria
hannelore.brandt@gmail.com

Åke Brännström
Evolution and Ecology Program,
International Institute for Applied
Systems Analysis,
Schlossplatz 1, 2361 Laxenburg,
Austria

Jingan Cui
Department of Mathematics,
Nanjing Normal UniversityNanjing
210097, China
cuija@njnu.edu.cn

J. M. Cushing
Department of Mathematics,
Interdisciplinary Program in Applied
Mathematics,
University of Arizona,
Tucson, Arizona 85721 USA
cushing@math.arizona.edu

Odo Diekmann
Department of Mathematics,
University of Utrecht, P.O. Box
80010, 3580 TA Utrecht,
The Netherlands
O.Diekmann@math.ruu.nl

Ulf Diekmann
Evolution and Ecology Program,
International Institute for Applied
Systems Analysis,
Schlossplatz 1, 2361 Laxenburg,
Austria
dieckmann@iiasa.ac.at

Mats Gyllenberg
Rolf Nevanlinna Institute Department of Mathematics and
Statistics,
FIN-00014 University of Helsinki,
Finland
mats.gyllenberg@helsinki.fi

Reinier HilleRisLambers
CSIRO Entomology, 120 Meiers
Road, Indooroopilly, QLD 4068,
Australia

Mudassar Imran
Arizona State University, Tempe,
Arizona, 85287 USA
imran@mathpost.asu.edu

Hiroshi C. Ito
Graduate School of Arts and
Sciences,
University of Tokyo,
3-8-1 Komaba, Meguro-ku,
Tokyo 153-8902, Japan

List of Contributors

Yoh Iwasa
Department of Biology,
Faculty of Sciences,
Kyushu University, Japan
yiwasscb@mbox.nc.kyushu-u.ac.jp

J.A.J. Metz
Evolutionary and Ecological
Sciences, Leiden University,
Kaiserstraat 63, NL-2311 GP Leiden,
The Netherlands and Adaptive
Dynamics Network, IIASA,
A-2361 Laxenburg, Austria
metz@rulsfb.leidenuniv.nl

Hisashi Ohtsuki
Department of Biology,
Faculty of Sciences,
Kyushu University, Japan
ohtsuki@bio-math10.biology.
kyushu-u.ac.jp

Kazunori Sato
Department of Systems Engineering,
Faculty of Engineering,
Shizuoka University, Japan
sato@sys.eng.shizuoka.ac.jp

Karl Sigmund
Fakultät für Mathematik,
Nordbergstrasse 15, 1090 Wien,
Austria
karl.sigmund@gmail.com

Hal L. Smith
Arizona State University, Tempe,
Arizona, 85287 USA
halsmith@asu.edu

Yasuhiro Takeuchi
Department of Systems Engineering,
Faculty of Engineering,
Shizuoka University, Japan
takeuchi@sys.eng.shizuoka.ac.jp

1
Ecology as a Modern Science

Kazunori Sato, Yoh Iwasa, and Yasuhiro Takeuchi

Mathematical or theoretical modeling has gained an important role in ecology, especially in recent decades. We tend to consider that various ecological phenomena appearing in each species are governed by general mechanisms that can be clearly or explicitly described using mathematical or theoretical models. When we make these models, we should keep in mind which charactersitics of the focal phenomena are specific to that species, and extract the essentials of these phenomena as simply as possible. In order to verify the validity of that modeling, we should make quantitative or qualitative comparisons to data obtained from field measurements or laboratory experiments and improve our models by adding elements or altering the assumptions. However, we need the foundation of mathematics on which the models are based, and we believe that developments both in modeling and in mathematics can contribute to the growth of this field.

In order for ecology to develop as a science we must establish a solid foundation for the modeling of population dynamics from the individual level (mechanistically) not from the population level (phenomenologically). One may compare this to the historical transformation from thermodynamics to statistical mechanics. The derivation of population dynamical modeling from individual behavior is sometimes called "first principles", and several kinds of population models are successfully derived in these schemes. The other kind of approaches is referred to as "physiologically structured population models", which gives the model description by i-state or p-state at the individual or the population level, respectively, and clarifies the relation between these levels. In the next chapter Diekmann et al. review the mathematical framework for general physiologically structured population models. Furthermore, we learn the association between these models and a dynamical system.

Behavioral ecology or social ecology is one of the main topics in ecology. In these study areas the condition or the characteristics for evolution of some kind of behavior is discussed. Evolutionarily stable strategy (ESS) in game theory is the traditional key notion for these analyses, and, for example, can help us to understand the reason for the evolution of altruism, which has been one of the biggest mysteries since Darwin's times, because it seems to be disadvantageous to the altruistic individuals at first glance. Reciprocal altruism may be considered as one of the most probable candidates for the evolution of altruism, which initially appears to cause the decrease of each individual's fitness with such behavior but an increase over a longer period, namely within his or her lifespan. Brandt et al. give an excellent review on indirect reciprocation and investigate the evolutionary stability for their model.

Classical population dynamics assumes that interactions such as competition or prey-predator between species are described by total densities of a whole population. However, it is natural to consider that these interactions occur on a local spatial scale, and the models incorporating space, sometimes called "spatial ecology", have been intensely studied recently. The metapopulation model is the most studied. It consists of many subpopulations with the risk of local extinction in each subpopulation and the recolonization by other subpopulations. Sometimes the metapopulation can persist longer than the single isolotated population because of the asynchronized dynamics between these subpopulations which is considered one of the important characteristics of metapopulation dynamics. We have recognized the usefulness of the metapopulation structure by the accumulating number of cases in which the metapopulation model seems to resemble the real ecological dynamics, especially concerning the local extinction and recolonization key concepts in the conservation of species (conservation biology). The simplest case of metapopulation corresponds to the two-patch structured models, and Cui & Takeuchi analyze the time dependent dispersal between these patches by non-autonomous equations with periodic functions or with dispersal time delays.

Lattice models are another kind of spatial model, in which individuals or subpopulations are regularly arranged in space and the interactions between them are restricted to neighbors. We also use the terms "interacting particle systems" or "cellular automata" when we categorize these models, depending on whether the dynamics is given in continuous or discrete time, respectively. Sato reviews the sexual reproduction process in which the mean-field approximation never corresponds to the fast stirring or diffusion, and utilizes the pair approximation, which is well known as a useful technique in the analysis of lattice models, to study the case without stirring for this model.

We need to consider ecological matters for a wide range of biological species (from bacteria to mammals), the various environments that are their habitats (soil, terrestrial, or aquatic) and the scale (from individual to ecosystem). We should take care to adopt the optimal modeling for each of these domains. The population dynamics of microorganisms can be most appropriately dealt with using deterministic differential equations. Imran & Smith analyze the population dynamics of bacteria with and without plasmids on biofilms.

Next we want to take an unsual interdiciplinary research project "Nonlinear Population Dynamics" which is a well known collaboration between experimentalists and mathematicians named "Beetles", dealing with flour beetles *Tribolium*. Cushing gives an excellent review of the results obtained by this project and leads us to recognize the importance of nonlinearity and stochasticity in population dynamics afresh.

In the final chapter, Dieckmann et al. explain in detail the notion of the adaptive dynamics theory with several examples. This is expected to become the model for understanding community structures by the linking of ecology and evolution. We learn how this theory analyzes the community structure in terms of stability, complexity or diversity, structure that is produced by the interaction of ecological communities and evolutionary processes.

In this volume readers will become familiar with various kinds of mathematical and theoretical modeling in ecology, and also techniques to analyze the models. They may find some treasures for the solution of their own present questions and new problems for the future. We believe that mathematical and theoretical analyses can be used to understand the corresponding ecological phenomena, but the models should if necessary be revised so that they coincide with field measurements or experimental data. Today's modern science of ecology integrates theories, models and data, all of which interact to continually improve our understanding.

2

Physiologically Structured Population Models: Towards a General Mathematical Theory

Odo Diekmann, Mats Gyllenberg, and Johan Metz

Summary. We review the state-of-the-art concerning a mathematical framework for general physiologically structured population models. When individual development is affected by the population density, such models lead to quasilinear equations. We show how to associate a dynamical system (defined on an infinite dimensional state space) to the model and how to determine the steady states. Concerning the principle of linearized stability, we offer a conjecture as well as some preliminary steps towards a proof.

2.1 Ecological motivation

How do phenomena at the population level (p-level) relate to mechanisms at the individual level (i-level)? When investigating the relationship, it is often necessary to distinguish individuals from one another according to certain physiological traits, such as body size and energy reserves. The resulting p-models are called "physiologically structured" (Metz and Diekmann 1986). They combine an i-level submodel for "maturation", i.e., change of i-state, with submodels for "survival" and "reproduction", which concern changes in the number of individuals. So they are "individual based", in the sense that the submodels apply to processes at the i-level. Yet they usually (but not necessarily) employ deterministic bookkeeping at the p-level (so they involve an implicit "law of large numbers" argument).

A first aim of this paper is to explain a systematic modelling approach for incorporating interaction. The key idea is to build a nonlinear model in two steps, by explicitly introducing, as step one, the environmental condition via the requirement that individuals are independent from one another (and hence equations are linear) when this condition is prescribed as a function of time. The second step then consists of modelling the feedback law that describes how the environmental condition depends on the current population size and composition.

Let us sketch three examples, while referring to de Roos and Persson (2001, 2002) and de Roos, Persson and McCauley (2003) for more details, additional examples and motivation as well as further references.

If juveniles turn adult (i.e., start reproducing) only upon reaching a certain size, there is a *variable maturation delay* between being born and reaching adulthood. Since small individuals need less energy for maintenance than large individuals, the juveniles can outcompete their parents by reducing the food level so much that adults starve to death. Thus "cohort cycles" may result, i.e., the population can consist of a cohort of individuals which are all born within a small time window. Once the cohort reaches the adult size it starts reproducing, thus producing the next cohort, but then quickly dies from starvation. So here the p-phenomenon is the occurrence of cohort cycles (which are indeed observed in fish populations in several lakes (Persson et al. 2000)) and the i-mechanism is the combination of a minimal adult size with a food concentration dependent i-growth rate.

The second example concerns *cannibalistic interaction*. Again we take i-size as the i-state, now since bigger individuals can eat smaller ones, but not vice versa. The p-phenomenon is that a population may persist at low renewal rates for adult food, simply since juvenile food becomes indirectly available to adults via cannibalism (the most extreme example is found in some lakes in which a predatory fish, such as pike or perch, occurs but no other fish whatsoever, cf. Persson et al. 2000, 2003). So reproduction becomes similar to farming, gaining a harvest from prior sowing (Getto, Diekmann and de Roos, submitted).

The third example is a bit more complex. It concerns the interplay between competition for food and mortality from predation in a size structured consumer population that is itself prey to an exploited (by humans) predator population, where the predators eat only small prey individuals. The phenomenon of interest is a bistability in the composition of the consumer population with severe consequences for the predators. At low mortality from predation, a large fraction of the consumers pass through the vulnerable size range, leading to a severe competition for food and a very small per capita as well as total reproductive output. The result is a consumer population consisting of stunted adults and few juveniles, a size structure that keeps the predators from (re-)entering the ecosystem. However, if the ecosystem is started up with a high predator density, due to a history in which parameters were different, these predators, by eating most of the young before they grow large, cause the survivors to thrive, with a consequent large total reproductive output. Thus, the predators keep the density of vulnerable prey sufficiently high for the predator population to persist. If exploitation lets the predator population diminish below a certain density, it collapses due to the attendant change in its food population.

Interestingly, a similar phenomenon can occur if the predators preferentially eat the larger sized individuals only. A more detailed analysis by de Roos, Persson and Thieme (2003) shows that the essence of the matter is that in the absence of predators the consumer population is regulated mainly by the rate at which individuals pass through a certain size range, with the predators specialising on a different size range. As noted by de Roos and Persson (2002), a mechanism of this sort may well explain the failure of the Northwest Atlantic cod to recover after its collapse from overfishing: After the cod collapsed, the abundance of their main food, capelin, increased, but capelin growth rates decreased and adults became significantly smaller. (See Scheffer et al. (2001) for a general survey on catastrophic collapses.)

A large part of this paper is based on earlier work of ours, viz. (Diekmann et al. 1998, 2001, 2003), which we shall refer to as Part I, Part II, and Part III, respectively. The reader is referred to (Ackleh and Ito, to appear; Calsina and Saldaña, 1997; Cushing, 1998; Tucker and Zimmermann, 1988) for alternative approaches.

2.2 Model ingredients for linear models

Let the i-state, which we shall often denote by the symbol x, take values in the i-state space Ω. Usually Ω will be a nice subset of $\mathbb{R}^k$ for some k. As an example, let $x = \begin{pmatrix} a \\ y \end{pmatrix}$ with a the age and y the size of an individual. Then Ω could be the positive quadrant $\{x \colon a \geq 0,\ y \geq 0\}$ or some subset of this quadrant.

We denote the environmental condition, either as a function of time or at a particular time, by the symbol I. In principle I at a particular time is a function of x, since the way individuals experience the world may very well be i-state specific. For technical reasons, we restrict our attention to environmental conditions that are fully characterized in terms of finitely many numbers (i.e., $I(t) \in \mathbb{R}^k$ for some k and x-dependence is incorporated via fixed weight functions as explained below by way of an example). The technical reasons are twofold. Firstly, this seems a necessary approximation when it comes to numerical solution methods. Secondly, as yet we have not developed any existence and uniqueness theory for the initial value problem in cases in which the environmental condition is i-state specific (and to do so one has to surmount substantial technical problems (Kirkilionis and Saldaña, in preparation).

As an example, think of $I = \begin{pmatrix} I_1 \\ I_2 \end{pmatrix}$, with I_1 the concentration of juvenile food and I_2 the concentration of adult food. We may then describe the food concentration as experienced by an individual of size y by the linear combination $\phi_1(y)I_1 + \phi_2(y)I_2$, where ϕ_1 is a decreasing function while ϕ_2 is increasing. Thus we can incorporate that the food preference is y-specific and gradually changes from juvenile to adult food.

The environmental condition should be chosen such that individuals are independent from one another when I is given as a function of time. The

i-state should be such that all information about the past of I, relevant for predicting future i-behaviour, is incorporated in the current value of the i-state. Here "i-behaviour" first of all refers to contribution to population changes, i.e., to death and reproduction (note that at the i-level this may very well amount to specifying probabilities per unit of time), but once the i-state has been introduced it also refers to predicting future i-states from the current i-state (possibly in the form of specifying a probability density).

As a notational convention we adopt that an environmental condition I is defined on a time interval $[0, \ell(I))$. Often we call I an *input* and $\ell(I)$ the *length* of the input. For $s \leq \ell(I)$ we then denote by $\rho(s)I$ the *restriction* of I to the interval $[0, s)$. By defining

$$(\theta(-s)I)(\tau) = I(\tau + s) \quad \text{for} \quad 0 \leq \tau < \ell(I) - s \tag{1}$$

we achieve that $\theta(-s)I$ incorporates the information about the restriction of I to $[s, \ell(I))$ but, by shifting back, in the form of a function defined on $[0, \ell(I) - s)$. We write

$$I = \theta(-s)I \odot \rho(s)I \tag{2}$$

where the symbol $\odot$ denotes *concatenation* defined by

$$(J \odot K)(\tau) = \begin{cases} K(\tau) & 0 \leq \tau < \ell(K) \\ J(\tau - \ell(K)) & \ell(K) \leq \tau < \ell(K) + \ell(J) \end{cases} \tag{3}$$

A *linear* structured population model is defined in terms of two ingredients, u and Λ, which are both functions of I, x and ω, where ω is a measurable subset of Ω (which thus implies the requirement that Ω comes equipped with a σ-algebra Σ). The interpretation is as follows:

$u_I(x, \omega)$ is the probability that, given the input I, an individual which has i-state $x \in \Omega$ at a certain time, is still alive $\ell(I)$ units of time later and then has i-state in $\omega \in \Sigma$;

$\Lambda_I(x, \omega)$ is the number of offspring, with state-at-birth in $\omega \in \Sigma$, that an individual is expected to produce when it gets exposed to the input I while starting in x, during the total length of the input.

This interpretation requires that certain consistency relations and monotonicity conditions should hold. In order to formulate these we first introduce some terminology and notation. We want u and Λ to be *parametrized positive kernels*, where I is the "parameter" and a kernel k is a map from $\Omega \times \Sigma$ into $\mathbb{R}$ which is bounded and measurable with respect to the first variable and countably additive with respect to the second variable. We call a kernel positive if it assumes non-negative values only. The *product* $k \times l$ of two kernels k and l is the kernel defined by

$$(k \times l)(x, \omega) = \int_\Omega k(\xi, \omega) l(x, d\xi). \tag{4}$$

Assumption 2.2.1

(i) *Chapman-Kolmogorov:*

$$u_{I \odot J} = u_I \times u_J \qquad (5)$$

(ii) *Reproduction-survival-maturation consistency:*

$$\Lambda_{I \odot J} = \Lambda_J + \Lambda_I \times v_J \qquad (6)$$

(iii) $\sigma \mapsto \Lambda_{\rho(\sigma)I}(x,\omega)$ *is non-decreasing with limit zero for $\sigma \downarrow 0$ (the monotonicity actually follows from (6) and positivity).*

(iv) $\sigma \mapsto u_{\rho(\sigma)I}(x,\Omega)$ *is non-increasing and*

$$\lim_{\sigma \downarrow 0} u_{\rho(\sigma)I}(x,\omega) = \delta_x(\omega) \,.$$

(v) *In addition we require finite life expectancy: there exists $M < \infty$ such that*

$$\int_{(0,\ell(I))} \sigma u_{\rho(d\sigma)I}(x,\Omega) \le M$$

for all $x \in \Omega$ and all I.

If maturation is deterministic, the ingredient u_I can be put into a particularly simple and useful form. Consider an individual with i-state x at a certain time. Let $X_I(x)$ be the i-state of that individual $\ell(I)$ units of time later, given the input I and let $\mathcal{F}_I(x)$ be its survival probability. Then

$$u_I(x,\omega) = \mathcal{F}_I(x) \delta_{X_I(x)}(\omega) \,. \qquad (7)$$

Concerning the specification of Λ, it makes first of all sense to introduce the set Ω_b of possible states-at-birth (cf. Part I, Definition 2.5; the idea is that $\Lambda_I(x,\omega) = 0$ whenever $\omega \cap \Omega_b = \emptyset$). Two situations are of special interest

- the discrete case: Ω_b is a finite set $\{x_{b_1}, x_{b_2}, \ldots, x_{b_m}\}$ (with the case $m = 1$ being of even stronger special interest)
- the absolutely continuous case: Ω_b is a lower dimensional manifold with a "natural" (Lebesgue) measure $d\xi$ defined on it, and $\Lambda_I(x,\cdot)$ is absolutely continuous with respect to that measure. Here the archetypical example is $\Omega_b = \{(a,x): a = 0, x_{\min} \le x \le x_{\max}\}$ that arises when modeling an age-size structured population.

In the case of a finite Ω_b we put

$$\Lambda_I(x,\omega) = \sum_{j=1}^{m} {}_j L_I(x) \delta_{x_{b_j}}(\omega) \qquad (8)$$

where ${}_j L_I(x)$ is the expected number of children, with i-state at birth x_{b_j}, produced, given the input I and in the period of length $\ell(I)$ of this input, by an individual having i-state x at the start of the input. In the case of Ω_b being a lower dimensional manifold we put

$$\Lambda_I(x,\omega) = \int_{\omega \cap \Omega_b} {}_\xi L_I(x)\,\mathrm{d}\xi, \tag{9}$$

where ${}_\xi L_I(x)$ has an analogous interpretation (but note that now it is a density with respect to ξ: only after integrating with respect to ξ over a subset of Ω_b do we get a number).

The building blocks X, $\mathcal{F}$ and L are, in turn, obtained as solutions of differential equations when the i-model is formulated in terms of a maturation rate g, a per capita death rate μ and a per capita (state-at-birth specific) reproduction rate β. These read

$$\begin{cases} \frac{\mathrm{d}}{\mathrm{d}t} X_{\rho(t)I}(x) = g(X_{\rho(t)I}(x), I(t)) \\ X_{\rho(0)I}(x) = x \end{cases} \tag{10}$$

$$\begin{cases} \frac{\mathrm{d}}{\mathrm{d}t} \mathcal{F}_{\rho(t)I}(x) = -\mu(X_{\rho(t)I}(x), I(t))\mathcal{F}_{\rho(t)I}(x) \\ \mathcal{F}_{\rho(0)I}(I) = 1 \end{cases} \tag{11}$$

$$\begin{cases} \frac{\mathrm{d}}{\mathrm{d}t} {}_\xi L_{\rho(t)I}(x) = \beta_\xi(X_{\rho(t)I}(x), I(t))\mathcal{F}_{\rho(t)I}(x) \\ {}_\xi L_{\rho(0)I}(x) = 0 \end{cases} \tag{12}$$

or, in short hand notation,

$$\begin{cases} \frac{\mathrm{d}X}{\mathrm{d}t} = g(X, I) \\ \frac{\mathrm{d}\mathcal{F}}{\mathrm{d}t} = -\mu(X, I)\mathcal{F} \\ \frac{\mathrm{d}L}{\mathrm{d}t} = \beta(X, I)\mathcal{F} \end{cases} \tag{13}$$

We conclude that the ingredients u and Λ for a linear structured population model can be constructively defined in terms of solutions $X, \mathcal{F}$ and L of ordinary differential equations involving the ingredients g, μ and β which specify the i-behaviour in terms of rates as a function of the current i-state and the prevailing environmental condition.

When i-state development is stochastic, rather than deterministic, one needs to replace (10). For instance, if i-state corresponds to spatial position and individuals perform Brownian motion, one needs to replace (10) by the diffusion equation for the probability density of finding the individual at a position after some time, given I. The advantage of "starting" from the ingredients u and Λ is that they encompass all such variations.

It is straightforward to check that, under appropriate assumptions on g, μ and β, (7)–(8)/(9) define parametrised positive kernels satisfying Assumption 2.2.1.

The true modelling consists of a specification of g, μ and β, see e.g. (Kooijman, 2000).

2.3 Feedback via the environmental condition

At any time t a population is described by a positive measure $m(t)$ on Ω. Possibly this measure is absolutely continuous (with respect to the Lebesgue measure; again we think of Ω as a subset of $\mathbb{R}^k$). Then there is a density function $n(t,\cdot)$, defined on Ω, such that

$$m(t)(\omega) = \int_\omega n(t,x)\,\mathrm{d}x\ . \tag{14}$$

To illustrate the idea of interaction via environmental variables, we consider the situation of competition for food. Let the dynamics of the substrate S be generated by

$$\frac{\mathrm{d}S}{\mathrm{d}t} = \frac{1}{\varepsilon}\left(S_0 - S - S\int_\Omega \gamma(x)m(t)(\mathrm{d}x)\right), \tag{15}$$

where $\varepsilon^{-1}\gamma$ is the i-state specific per capita consumption rate. So an individual with i-state x ingests $\varepsilon^{-1}\gamma(x)S$ units of substrate per unit of time. In energy budget models (Kooijman, 2000) one often assumes that a fraction $1 - \kappa(x)$ of the ingested energy is scheduled to growth and maintenance and the remaining fraction $\kappa(x)$ to reproduction. Thus the $\varepsilon^{-1}\gamma(x)S$ enters in the specification of g and β (and, in case of starvation, i.e., when maintenance cannot be covered, also μ). So the S is (a component of) I. Vice versa, the factor $\int_\Omega \gamma(x)m(t)(\mathrm{d}x)$ corresponds to the environmental condition for the substrate population. It appears that we can couple the substrate and the consumer population via the idea that one constitutes the environmental condition for the other.

If the time scale parameter ε in (15) is very small one can employ the quasi-steady-state approximation for the substrate, i.e., require that the factor within brackets at the right hand side of (15) equals zero. This yields

$$S = \frac{S_0}{1+I} \tag{16}$$

where

$$I(t) = \int_\Omega \gamma(x)m(t)(\mathrm{d}x) \tag{17}$$

One should interpret these two identities as follows. When I is considered as given, as an *input*, the formula (16) specifies what substrate density the individuals of the consumer population experience. And this then in turn determines how the I enters the expressions for g, β and, possibly, μ. The identity (17), on the other hand, is the feedback law specifying how, in fact, the I at a particular time relates to the extant population at that time. In other words, the combination of (16) with rules for how g, β and μ depend on S defines a linear structured population model. But if we add to that the consistency requirement (17) we turn the linear model into a nonlinear model in which it is incorporated that individuals interact by competing for a limited resource S. Note that the ingredients g, μ and β of the linear model need to be supplemented by the ingredient γ in order to define the nonlinear model. One could call $\gamma(x)$ the i-state specific contribution to the environmental condition. (The precise interpretation depends on the meaning of (the component of) I).

Since the environmental condition is chosen such that individuals are, for given I, independent of one another, the feedback law (17) is necessarily linear. Or, phrased differently, the components of I are linear functionals of the p-state. We call (17) a *pure mass-action* feedback law.

Sometimes the specification of g, μ and β is based on submodels for behavioural processes at a very short time scale, the most well-known example being the Holling type II functional response as derived from a submodel in which predators can be either searching for prey or busy handling prey that has been caught. In such cases the feedback law exhibits a certain hierarchical structure which is described in Part II, Sect. 6 and which we have called *generalized mass action*. In this paper we restrict ourselves to the pure mass action case (17).

Especially in the modeling phase it is often helpful to close the feedback loop in two steps: first an *output* is computed, which then is fed back as input via a feedback map. In the example considered above we would write (17) as

$$O(t) := \int_\Omega \gamma(x) m(t)(\mathrm{d}x) \,. \tag{18}$$

Considering S as the true input, we would then write (16) as

$$S = F(O) \,, \tag{19}$$

where

$$F(O) = \frac{S_0}{1+O} \,. \tag{20}$$

The advantage is twofold: i) it represents better what is going on biologically, and ii) one can use (18) as a definition, with (19) as the equation that closes the feedback loop. In contrast (16), by combining both steps in one, lacks such a clear interpretation. From a mathematical viewpoint the role of (17) is that of an equation only, while the modelling aspect, i.e., the definition of what inputs and outputs amount to observationally, is lost from sight. On the other hand, the drawback of distinguishing between I and O is that an additional variable is introduced which clutters the analysis without playing any useful role. So in the following we use only I.

2.4 Construction of p-state evolution. Step 1: the linear case.

For the sake of exposition we restrict ourselves here to the situation of a fixed state-at-birth x_b. Given an initial p-state m, we define the cumulative first generation offspring function B^1 by

$$B^1(t) = \int_\Omega L_{\rho(t)I}(x) m(\mathrm{d}x) \,. \tag{21}$$

The cumulative second generation offspring function B^2 is next defined by

$$B^2(t) = \int_0^t L_{\rho(t-\tau)\theta(-\tau)I}(x_b) B^1(\mathrm{d}\tau) \,, \tag{22}$$

et cetera (that is, replace in (22) B^2 by B^{n+1} and B^1 by B^n). The cumulative "all offspring" function

$$B^c = \sum_{n=1}^\infty B^n \tag{23}$$

then satisfies the renewal equation

$$B^c(t) = B^1(t) + \int_0^t L_{\rho(t-\tau)\theta(-\tau)I}(x_b) B^c(\mathrm{d}\tau) \tag{24}$$

and one can view (23) as the generation expansion obtained by solving (24) by successive approximation. Note that B^c depends on I, even though we do not incorporate this in the notation.

If we denote by $T_I m$ the p-state at time $\ell(I)$, given that the p-state at time zero is m and given the time course I of the environmental condition, then

$$T_I m = u_I \times m + \int_0^{\ell(I)} u_{\theta(-\tau)I}(x_b, \cdot) B^c(\mathrm{d}\tau) \tag{25}$$

where

$$(u_I \times m)(\omega) = \int_\Omega u_I(x,\omega) m(\mathrm{d}x) \tag{26}$$

describes the survival and maturation of the individuals present at time zero, while the second term takes into account the survival and maturation of all individuals born after time zero. The key result of Part I is that the operators T_I form a semigroup, that is, the map $I \mapsto T_I$ transforms concatenation (recall (2)) into composition of maps:

Theorem 2.4.1
$$T_I = T_{\theta(-t)I} T_{\rho(t)I}$$

for any $t \in [0, l(I))$

Let us recapitulate. Starting from g, μ and β, one constructs u and L (recall (8)); if there is only one possible state-at-birth, then Λ is completely determined by L). Given an initial p-state m one next constructively defines the solution B^c of (24) by (23). The formula (25) then provides a way to calculate, given I, the p-state after $\ell(I)$ units of time from u, B^c and m. And Theorem 2.4.1 justifies our use of the word "p-state": our construction yields a dynamical system.

Even though we rightfully refer to Part I for Theorem 2.4.1, readers who want to see more details are advised to first consult Part II since some of our current notation goes back only to that reference.

2.5 Construction of p-state evolution.
Step 2: closing the feedback loop.

If we substitute $m(t) = T_{\rho(t)I}m$ into (17) we obtain the equation

$$I(t) = \gamma \times T_{\rho(t)I}m = \int_\Omega \gamma(x)(T_{\rho(t)I}m)(\mathrm{d}x) \qquad (27)$$

that I should satisfy in order to have consistency between input and output. We view (27) as a fixed point problem for I, parametrised by the initial p-state m.

In Sects. 7 and 8 of Part II one finds various assumptions on u, Λ and γ, respectively, g, μ, β and γ that guarantee that the right hand side of (27) defines a contraction mapping on a suitable function space. Here "suitable" in particular involves a restriction for the length l of the interval on which I is defined. Thus the contraction mapping principle yields a local solution $I = I_m$ of (27). One next notes (see Diekmann and Getto, to appear, for details) that:

- a fixed point on a smaller interval is a restriction of a fixed point on a larger interval,
- $\theta(-t)I_m = I_{T_{\rho(t)I_m}m}$, roughly saying that shifted fixed points are the fixed points corresponding to the updated p-state,
- uniqueness holds on any interval,
- fixed points can be concatenated to achieve continuation, that is, to obtain solutions on longer time intervals

to conclude that the local solution can be extended to a maximal solution, which we also denote by I_m. A key result of Part II is that the definition

$$S(t, m) = T_{\rho(t)I_m}m \qquad (28)$$

yields a semiflow:

Theorem 2.5.1
$$S(t+s, m) = S(t, S(s, m))$$

(Again we refer to Diekmann & Getto, to appear, for details and for various results about boundedness and global existence as well as weak*- continuity with respect to time t and initial condition m.)

2.6 Steady states

The symbol $\bar{I}$ denotes a constant input defined on $[0, \infty)$. (Slightly abusing notation we do not distinguish between the function and the value it takes.) A *steady state* is a measure $\bar{m}$ on Ω such that

$$T_{\rho(t)\bar{I}}\bar{m} = \bar{m}, \quad \forall t \geq 0, \tag{29}$$

where

$$\bar{I} = \gamma \times \bar{m} = \int_\Omega \gamma(x)\bar{m}(\mathrm{d}x). \tag{30}$$

Since $\widetilde{T}(t) := T_{\rho(t)\bar{I}}$ is a semigroup of *positive* linear operators and $\bar{m}$ has to be positive, (29) amounts to the condition that the spectral radius is an eigenvalue and is equal to one. (For future reference we observe that, whenever there is a spectral gap,

$$\widetilde{T}(t)m \to c\bar{m} \quad \text{as} \quad t \to \infty$$

exponentially in the weak*-sense, for any positive initial measure m. Here $c = c(m)$ is a positive real number.)

The defining relations (29)–(30) are not suitable for "finding" steady states. For that purpose, the generation perspective is much more suitable. In particular one can concentrate on newborn individuals and the offspring they are expected to produce, with due attention to the state-at-birth of the offspring.

In the simple case of one possible state-at-birth, a first steady state condition is that the basic reproduction ratio, the expected number of offspring, equals one:

$$R_0(\bar{I}) := L_{\rho(\infty)\bar{I}}(x_\mathrm{b}) = 1 \tag{31}$$

This is a condition on $\bar{I}$. If $\dim I = 1$ this is one equation in one unknown. Very often R_0 is monotone in $\bar{I}$ which then immediately yields uniqueness.

More generally we should, in the notation of (26), have

$$\Lambda_{\rho(\infty)\bar{I}} \times b = b, \tag{32}$$

with b a positive measure on the set Ω_b of possible birth states. Written out in detail (32) reads

$$\int_{\Omega_\mathrm{b}} \Lambda_{\rho(\infty)\bar{I}}(x, \omega)b(\mathrm{d}x) = b(\omega) \tag{33}$$

for all measurable subsets ω of Ω_b. And if Ω_b is a nice subset of $\mathbb{R}^k$ for some k and b has a density f we may rewrite this as

$$\int_{\Omega_b} \varepsilon L_{\rho(\infty)\bar{I}}(x) f(x)\,\mathrm{d}x = f(\xi)\,,\quad \xi \in \Omega_b\,. \tag{34}$$

Equation (32) is a linear eigenvalue problem: the dominant eigenvalue of a positive operator should be one. This is, just as (31) but now more implicitly, a condition on the parameter $\bar{I}$. If this condition is satisfied and the eigenvalue is algebraically simple (a sufficient condition being the irreducibility of the positive operator) then the eigenvector b is determined uniquely modulo a positive multiplicative constant, to be denoted by c below.

Returning to the case of a fixed state-at-birth, we note that (10)–(12) simplify considerably when the input is constant. For given $\bar{I}$ we define $\bar{x}$ and $\bar{\mathcal{F}}$ by

$$\begin{cases} \frac{\mathrm{d}\bar{x}}{\mathrm{d}a} = g(\bar{x},\bar{I}) \\ \bar{x}(0) = x_b \end{cases} \tag{35}$$

$$\begin{cases} \frac{\mathrm{d}\bar{\mathcal{F}}}{\mathrm{d}a} = -\mu(\bar{x},\bar{I})\bar{\mathcal{F}} \\ \bar{\mathcal{F}}(0) = 1 \end{cases} \tag{36}$$

and next we note that

$$R_0(\bar{I}) = \int_0^\infty \beta(\bar{x}(a),\bar{I})\bar{\mathcal{F}}(a)\,\mathrm{d}a\,. \tag{37}$$

Let c denote the steady p-birth rate. Then

$$\bar{m}(\omega) = c\int_0^\infty u_{\rho(a)\bar{I}}(x_b,\omega)\,\mathrm{d}a = c\int_0^\infty \bar{\mathcal{F}}(a)\delta_{\bar{x}(a)}(\omega)\,\mathrm{d}a \tag{38}$$

and consequently (30) can be written as

$$\bar{I} = c\int_0^\infty \bar{\mathcal{F}}(a)\gamma(\bar{x}(a))\,\mathrm{d}a\,. \tag{39}$$

Beware that $\bar{\mathcal{F}}$ and $\bar{x}$ depend on $\bar{I}$.

Theorem 2.6.1 $\bar{m}$ *is a steady state, i.e., (29)–(30) hold, iff $\bar{m}$ is given by (38), with $\bar{x}$ and $\bar{\mathcal{F}}$ defined by (35)–(36), where $\bar{I}$ and c are such that (31) (with $R_0(\bar{I})$ given by (37)) and (39) hold.*

For the proof see Part III. Note that (31) and (39) are $1+\dim I$ equations in as many unknowns, viz., c and $\bar{I}$. Also note that (37) is defined completely in terms of solutions of ODE, since we may supplement (35)–(36) with

$$\begin{cases} \frac{\mathrm{d}\bar{L}}{\mathrm{d}a} = \beta(\bar{x},\bar{I})\bar{\mathcal{F}} \\ \bar{L}(0) = 0 \end{cases} \tag{40}$$

and put
$$R_0(\overline{I}) = \overline{L}(\infty) \,. \tag{41}$$
Similarly we may write (39) as
$$\overline{I} = cG(\infty) \tag{42}$$
where G is obtained by solving
$$\begin{cases} \frac{dG}{da} = \gamma(\overline{x})\overline{\mathcal{F}} \\ G(0) = 0 \end{cases} \,. \tag{43}$$

The main message of Kirkilionis et al. (2001) is that one can do a numerical parameter continuation study of steady states of physiologically structured population problems by combining standard ODE solvers with standard continuation algorithms when solving (31)–(39).

2.7 Linearized stability

Given a steady state, how do we determine whether or not it is stable? Apart from the special situation in which we want to determine the ability of a missing species to invade successfully an existing community (see e.g. Part III, Sects. 2 and 3 where it is explained that the answer can be given in terms of R_0), this is a difficult question. We say that the answer can be found by way of a *characteristic equation* if it is possible to derive a function $f \colon \mathbb{C} \to \mathbb{C}$ such that the steady state is asymptotically stable if all roots of the equation $f(\lambda) = 0$ lie in the left half plane while being unstable if at least one root lies in the right half plane. We claim that for physiologically structured population models the answer can indeed be found by way of a characteristic equation and that, moreover, this equation takes the form
$$\det M(\lambda) = 0 \,, \tag{44}$$
where M is a $\dim I \times \dim I$ matrix. The intuitive explanation is that the semigroup $\tilde{T}(t) = T_{\rho(t)\overline{I}}$ of positive linear operators introduced in the beginning of Sect. 2.6 has dominant eigenvalue zero. Accordingly, the stability or instability is completely determined by the feedback loop (and not by the population dynamics *per se*) and this leads, after linearization, to a transcendental characteristic equation in terms of a matrix of size $\dim I \times \dim I$ (essentially the λ comes in via the Laplace transform of a time kernel; see below).

The proof of this claim is involved and, in fact, some details still have to be filled in. For the stability part there are two steps:

Step 1 assuming that $I_m(t) - \overline{I} \to 0$ exponentially for $t \to \infty$, show that $S(t, m) \to \overline{m}$ for $t \to \infty$,

Step 2 assuming that all roots of (44) are in the left half plane, show that $m - \overline{m}$ small implies that $I_m(t) - \overline{I} \to 0$ for $t \to \infty$ (in fact exponentially).

As usual, the instability part is more difficult; it was proved, for age structured models, by Prüß(1983); see also (Desch and Schappacher 1986; Clément et al. 1987). The difficulty is substantially enhanced in the present case by the fact that the nonlinear semigroup is *not* differentiable (indeed, there is a problem with, e. g., slightly shifted Dirac easures). Our "escape strategy" is to consider an invariant and attracting subset of the p-state space on which we have more smoothness. In work in progress, mainly by Philipp Getto, we use a different p-state representation to characterize this subset, viz., we use the history of I and the history of the population birth rate to identify the p-state. In our further description below we restrict our attention to the stability part.

The only nonlinear feature in the constructive definition of the semiflow S is the fixed point problem (27) for the environmental variable I. So that is the problem we should linearize. As a preparatory step we rewrite (27) in the form

$$I(t) - \overline{I} = \gamma \times (T_{\rho(t)I} - T_{\rho(t)\overline{I}})\overline{m} + \gamma \times T_{\rho(t)I}(m - \overline{m}) \qquad (45)$$

and introduce the map Q that describes how the output depends on the perturbation of the steady input

$$(QJ)(t) = \gamma \times (T_{\rho(t)(\overline{I}+J)} - T_{\rho(t)\overline{I}})\overline{m} . \qquad (46)$$

Now think of J as extended by zero for negative arguments and define the shift operator $\theta(s)$ by

$$(\theta(s)J)(\tau) = J(\tau - s) . \qquad (47)$$

Proposition 2.7.1 Q *is translation invariant:*

$$Q\theta(s) = \theta(s)Q . \qquad (48)$$

Even though the map $J \mapsto T_{\rho(t)(\overline{I}+J)}\overline{m}$ is in general not smooth, the map Q may very well be, as it involves the pairing with γ. We state this as an assumption.

Assumption 2.7.2 Q *is differentiable with derivative L.*

This is basically a smoothness assumption on γ. Admittedly the assumption is stated rather imprecisely, as we have not specified the function space of inputs. The idea, however, is to compute the derivative for any fixed t and to use the outcome to define a linear input-output map L.

Now observe that L inherits the translation invariance of Q and recall that "linear + translation invariant $\Rightarrow$ convolution" whence we have

Proposition 2.7.3 $(LJ)(t) = \int_0^t k(t - \tau)J(\tau)d\tau$ *for some kernel k.*

Finally, we define $M(\lambda)$ to be the Laplace transform of k minus the identity. In fact one can express k, and hence M, explicitly in terms of solutions of linearized ODE like, when one linearizes (10)

$$\frac{dY}{da} = \frac{\partial g}{\partial x}Y + \frac{\partial g}{\partial I}J. \tag{49}$$

We refer to Kirkilionis et al. (2001) for the details. Note that this characterization of k allows a numerical implementation. Thus, despite all the complications, one can make the linearized stability test operational in the context of concrete examples!

References

1. Ackleh, A. S. and Ito, K. (to appear). Measure-valued solutions for a hierarchically size-structured population *SIAM J. Appl. Math.*
2. Calsina, À. and Saldaña, J. (1997). Asymptotic behaviour of a model of hierarchically structured population dynamics, *J. Math. Biol.* 35:967–987.
3. Clément, P., Diekmann, O., Gyllenberg, M., Heijmans, H.J.A.M., Thieme, H.R. (1989). Perturbation theory for dual semigroups III. Nonlinear Lipschitz continuous perturbations in the sun reflexive case. In *Volterra integro-differential equations in Banach spaces and applications*, Trento 1987, G. Da Prato and M. Iannelli (Eds.), Pitman research Notes in Mathematics Series, 190, pp. 67–89.
4. Cushing, J.M. (1998). *An introduction to structured population dynamics*, CBMS-NSF Regional conference series in applied mathematics 71, SIAM, Philadelphia.
5. Desch and Schappacher (1986). Linearized stability for nonlinear semigroups. In *Differential Equations in Banach Spaces* (A. Favini and E. Obrecht, Eds.) Spinger Lecture Notes in Mathematics 1223, pp. 61–73.
6. Diekmann, O. and Getto, Ph. (to appear). Boundedness, global existence and continuous dependence for nonlinear dynamical systems describing physiologically structured populations, *Journal of Differential Equations*.
7. Diekmann, O., Gyllenberg, M., Metz, J.A.J., and Thieme, H.R. (1998). On the formulation and analysis of general deterministic structured population models: I Linear theory. *Journal of Mathematical Biology* 36: 349–388.
8. Diekmann, O., Gyllenberg, M., Huang, H., Kirkilionis, M., Metz, J.A.J. and Thieme, H.R. (2001). On the Formulation and Analysis of General Deterministic Structured Population Models. II. Nonlinear Theory. *Journal of Mathematical Biology* 43: 157–189.
9. Diekmann, O., Gyllenberg, M. and Metz, J.A.J. (2003). Steady-state analysis of structured population models, *Theoretical Population Biology* 63: 309–338.
10. Getto, Ph., Diekmann, O. and de Roos A.M. (submitted). On the (dis)advantages of cannibalism, submitted to *Journal of Mathematical Biology*.
11. Kooijman, S.A.L.M. (2000). *Dynamic Energy and Mass Budgets in Biological Systems*, Cambridge University Press, Cambridge.
12. Kirkilionis, M. and Saldaña, J. (in preparation). A height-structured forest model. http://www.iwr.uni-heidelberg.de/sfb/Preprints2001.html

13. Kirkilionis, M., Diekmann, O., Lisser, B., Nool, M., de Roos, A.M., and Sommeijer, B. (2001). Numerical continuation of equilibria of physiologically structured population models. I. Theory. Mathematical Models and Methods in Applied Sciences 11: 1101–1127.
14. Kirkilionis et al. (2001).
15. Metz, J.A.J. and Diekmann, O. (1986). *The Dynamics of Physiologically Structured Populations*. Lecture Notes in Biomathematics 68. Springer, Berlin.
16. Persson, L., Byström, P., and Wahlström, E. (2000). Cannibalism and competition in Eurasian perch: Population dynamics of an ontogenetic omnivore, *Ecology* 81: 1058–1071.
17. Persson, L., De Roos, A.M., Claessen, D., Byström, P., Lövgren, J., Sjögren, S., Svanbäck, R., Wahlström, E., and Westman, E. (2003). Gigantic cannibals driving a whole-lake trophic cascade, *PNAS* 100: 4035–4039
18. Prüß, J. (1983). Stability analysis for equilibria in age-specific population dynamics, *Nonl. Anal. TMA* 7: 1291–1313.
19. de Roos, A.M., Person, L. and Thieme, H.R. (2003). Emergent Allee effects in top predators feeding on structured prey populations, *Proc. R. Soc. Lond. B* 270: 611–618.
20. de Roos, A.M. and Persson, L. (2001). Physiologically structured models – from versatile technique to ecological theory, *Oikos* 94: 51–71.
21. de Roos, A.M. and Persson, L. (2002). Size-dependent life-history traits promote catastrophic collapses of top predators, *Proc. Natl. Acad. Sci. USA* 99: 12907–12912
22. de Roos, A.M., Persson, L. and McCauley, E. (2003). The influence of size-dependent life history traits on the structure and dynamics of populations and communities. *Ecol. Lett.* 6: 473–487.
23. Scheffer, M., Carpenter, S.R. , Foley, J.A., Folke, C. and Walker, B. (2001). Catastrophic shifts in ecosystems, *Nature* 413: 591–596.
24. Tucker and Zimmermann (1988). A nonlinear model of population dynamics containing an arbitrary number of continuous structure variables, *SIAM J. Appl. Math.* 48: 549–591.
25. Webb, G.F. (1985) *Nonlinear Age-Dependent Population Dynamics*, Marcel Dekker, New York.

3
A Survey of Indirect Reciprocity

Hannelore Brandt, Hisashi Ohtsuki, Yoh Iwasa, and Karl Sigmund

Summary. This survey deals with indirect reciprocity, i. e. with the possibility that altruistic acts are returned, not by the recipient, but by a third party. After briefly sketching how this question is dealt with in classical game theory, we turn to models from evolutionary game theory. We describe recent work on the assessment of interactions, and the evolutionary stability of strategies for indirect reciprocation. All stable strategies (the 'leading eight') distinguish between justified and non-justified defections, and therefore are based on non-costly punishment. Next we consider the replicator dynamics of populations consisting of defectors, discriminators and undiscriminating altruists. We stress that errors can destabilise cooperation for strategies not distinguishing justified from unjustified defections, but that a fixed number of rounds, or the assumption of an individual's social network growing with age, can lead to cooperation based on a stable mixture of undiscriminating altruists and of discriminators who do not distinguish between justified and unjustified defection. We describe previous work using agent-based simulations for 'binary score' and 'full score' models. Finally, we survey the recent results on experiments with the indirect reciprocation game.

3.1 Introduction

In evolutionary biology, the two major approaches to the emergence of cooperation are kin-selection, on one hand, and reciprocation, on the other. The latter, which is essential for understanding cooperation between non-related individuals and very prominent in human societies, can be subdivided into two parts of unequal size. In direct reciprocity, it is the recipient of a helpful action who eventually returns the aid. In indirect reciprocity, the return is provided by a third party. This possibility has originally been named 'third-party altruism' or 'generalised reciprocity' by Trivers (1971). Later, Alexander (1987) explored it under the (now common) heading of 'indirect reciprocity', see also Ferrière (1998) and Wedekind (1998). Indirect reciprocity is much less well studied than direct reciprocity, and offers interesting theoretical challenges.

Several mechanisms for indirect reciprocity are conceivable. It could be, for instance, that a person having been helped is enclined to help a third party in turn. In cyclical networks, this provides a plausible feedback loop. But studies by Boyd and Richerson (1989) and van der Heijden (1996) suggest that such networks have to be rather small and rigid.

Alexander suggested, in contrast, that indirect reciprocation is based on reputation and status. By giving help to others, individuals acquire a high reputation. If help is directed preferentially towards recipients with a high reputation, defectors will be penalised. Such indirect reciprocation based on reputation and status is the topic of this paper.

The two main reasons why reputation mechanisms are interesting show up at two stages in human evolution which could not be further apart. On the one hand, status and reputation may well have played a major role in the evolution of moral systems since the dawn of prehistory, boosting cooperation between non-relatives (a major cause for the evolutionary success of hominids) and possibly providing a major selective impetus for the emergence of language, as a means of transmitting information about group members through gossip (Alexander, 1987, Nowak and Sigmund, 1998a, Panchanathan and Boyd, 2003). On the other hand, the very recent advent of e-commerce makes the efficient assessment of reputations and moral hazard in trust-based transactions a burning issue. Anonymous one-shot interactions in global markets, rather than long-lasting repeated interactions through direct reciprocation, seem to play an ever-increasing role in today's economy (Bolton et al, 2002, Keser, 2002, Dellarocas, 2003).

The aim of this paper is to provide a survey of the model-based theoretical investigations of the concept of indirect reciprocation, and of the remarkable results on experimental economic games inspired by them.

3.2 Indirect reciprocation for rational players

Before approaching the subject in the spirit of evolutionary game dynamics, we should stress that the same topic can also be addressed within classical game theory. At a first glance, it may almost look like a non-issue in this context. Indeed, it is easy to see that the main classical results on repeated games survive unharmed if the single co-player with whom one interacts in direct reciprocation is replaced by the wider cast of co-players showing up in indirect reciprocation. This holds, in particular, for the folk theorem on repeated games. It states, essentially, that every feasible payoff larger than the maximin level which players can guarantee for themselves is obtainable by strategies in Nash equilibrium, provided that the probability for another round is sufficiently large (Fudenberg and Maskin, 1986, Binmore, 1992). This can be achieved, in particular, by 'trigger strategies' that switch to defection after the first defection of the co-player: for in that case, it makes no sense to exploit the co-player in one round, thereby forfeiting all chances for mutual cooperation in further rounds. Exactly the same argument holds for indirect reciprocation in a population where players are randomly matched between rounds, if they know the case-history of every co-player which they encounter, and refuse help to any individual who ever refused to help someone (Rosenthal 1979; Okuno-Fujiwara and Postlewait 1989; Kandori 1992). The difference between personal enforcement, in the former case, and community enforcement, in the latter, is irrelevant to the sequence of payoffs encountered by an individual player.

It must be noted, however, that with such trigger strategies, the defection of a single player A results in the eventual punishment of all players, and the breakdown of cooperation in the whole population. Indeed, if A defects in a given round, then the next player B who is asked to help A will refuse, and so will C when asked to help B, etc, so that defection spreads rapidly through the population. If the population consists of rational agents, player A will not defect. But if even one player fails to be rational, the whole community is under threat.

As Sugden (1986) suggested, this can be remedied by another trigger strategy, which distinguishes between justified and unjustified defections. Such a strategy is based on the notion of *standing*. Each individual has originally a good standing, and loses this only by refusing help to an individual in good standing. Individuals refusing help to someone in bad standing do not lose their good standing. In this way, cooperation can be channelled towards those who cooperate.

So far, so obvious. The situation becomes more interesting if one assumes that players have only a limited knowledge of their co-players past, or must cope with unintended defections caused, for instance, by an error, or by the lack of adequate ressources to provide the required help. Kandori (1992) seems to have been the first to study the effects of limited observability in this context. In the extreme case, players know only their own history. Kandori has shown that under certain conditions a so-called 'contagious' equilibrium can still ensure cooperation among rational players: the strategy consists in switching to defection after having encountered the first defection. A single defection by one player is 'signalled', in this sense, to the whole community: but the retaliation may reach the wrong-doer only after many rounds, creating havoc among innocents. Moreover, Kandori has shown that with random matching and no information processing, cooperation cannot be sustained if the population is sufficiently large. Interestingly, Ellison (1994) has shown that cooperation can be resumed, eventually, if such 'contagious' punishments stop after a signal defined by a public random variable. He notes, however, that such cooperative equilibria are very dependent on the assumption that all players are rational. On the other hand, Kandori (1992) has shown, that decentralised mechanisms of local information processing based on a label carried by each agent may allow simple equilibrium strategies leading to cooperation even if occasionally errors occur. After a unilateral defection, players must 'repent' by cooperating, while meekly accepting the defection of their co-players for a certain number of rounds.

3.3 Indirect reciprocation for evolutionary games

In evolutionary games, it is no longer possible to postulate that players settle on an equilibrium which is sustained by their anticipation of the payoff obtained when they deviate unilaterally. Players are not assumed to be rational, or able to think ahead, deliberate, or coordinate. Strategies are simple behavioral programs; they are supposed to spread within the population if they are successful in the sense of yielding a high payoff (see e.g. Hofbauer and Sigmund, 1998). Typically, one assumes that such strategies arise randomly within a small minority of the population, by mutation or some other process. The question then becomes whether simple trial-and-error mechanisms resembling natural selection are able to lead, in the long run, to the emergence of cooperative behaviour.

The first papers in this field, by Nowak and Sigmund (1998a,b), led to a number of theoretical and experimental investigations. Roughly speaking, by now the fact that cooperative behaviour based on indirect reciprocity can emerge through evolutionary mechanisms is no longer in doubt, but there is debate on which strategy it is most likely to be based.

In the evolutionary version of the indirect reciprocity game, one considers populations of players which are endowed with some simple strategies. Whenever two players meet in one round of that game, one of them is randomly assigned the role of the donor and the other the role of the recipient. The donor can give help to the recipient: in this case, the recipient's payoff increases by a benefit b whereas the donor's payoff decreases by $-c$, the cost of giving (with $c < b$). The donor can, alternatively, refuse to help, in which case the payoffs of both players are not affected. A player's strategy specifies under which conditions the player should give help, when in the role of the donor.

From time to time, players leave the population and are replaced by new players. The probability that a new player inherits a given strategy occuring within the population is proportional to its frequency, and to the average payoff achieved by players using this strategy. This mimicks selection, but it can just as well be interpreted as a learning process: in that case, players switch their strategies without actually having to die. Some models of evolutionary games also incorporate mutations, which introduce small numbers of players using strategies which were not present in the resident population.

The first model by Nowak and Sigmund (1998a) was based on the concept of a score, a numerical value for reputation. A player's score, at any given time, is defined as difference between the number of decisions to give help, and the number of decisions to refuse help, up to that time. The score of a player entering the game is zero: it then increases or decreases by one point in each round in which the player is in the position of a donor. The range of the score is the set of all integers. This is called the 'full score' model. In a second, 'binary' model, discussed in Nowak and Sigmund (1998b), the range is reduced to two numbers only, 0 (*bad*) or 1 (*good*). This reflects only the players' behaviour in their previous round as a donor. One can, of course, conceive many other ways for keeping score: for instance, by considering neither all the previous actions of the players, nor their last action only, but their last five or ten actions, etc. The decision whether to give help or not should then be based on the scores of the players involved. In particular, a recipient with a high score should be more likely to receive help.

3.4 Assessment and reprobation

So far, the length of the memory is an aspect which has not attracted much attention. Most of the debate has concentrated on another issue: how should the score be updated? The basic issue is the same as in the framework of games between rational players. Cooperation cannot be sustained without discriminating against defectors. Players who discriminate must, on occasion, refuse help. If this lowers their score, they will be discriminated against, in future encounters, and obtain a lower payoff. How can such strategies be selected?

One solution is almost obvious. It is to use the same distinction between justified and non-justified defections as Sugden, and hence to rely on the notion of standing. As Nowak and Sigmund (1998b) described it, 'a player is born with good standing, and keeps it as long as he helps players who are in good standing. Such a player can therefore keep his good standing even when he defects, as long as the defection is directed at a player with bad standing. We believe that Sugden's strategy is a good approximation to how indirect reciprocation actually works in human societies.' And to the question of Fehr and Fischbacher (2003): 'Should an individual who does not help a person with a bad reputation lose his good reputation?', the answer is, clearly no.

However, two aspects make it worthwhile to investigate image-scoring more closely: one is the argument that standing is a rather complex notion, and seems to require a constant monitoring of the whole population which may overtax the players. Suppose your recipient A has refused help to a recipient B in a previous round. Was this refusal justified? Certainly so, if B has proved to be a helper. But what if B has refused help to some C? Then you would have to know whether B's defection towards C was justified, etc. With direct reciprocation, you have only to keep track of your previous interactions with B. Even here, an error in perception can lead to a deadlock: it may happen that both players believe that they are in good standing and keep punishing each other in good faith (see Boerlijst et al. 1997). With indirect reciprocation the problem becomes much more severe: you have to keep track, not only of the antecedents of your current recipient, but of the past actions of the recipient's former recipients etc.

The second interesting aspect of scoring is related to the concept of costly punishment. It is easy to see that the threat of punishment can keep players on the path of cooperation, and thus can solve the social dilemma, which is resumed in the question: why do players contribute to a public good, instead of just exploiting it? They may simply do it to avoid punishment. But if punishment is costly to the punisher, a 'second order social dilemma' arises: why should players shoulder the burden of punishing others? The doctrine of strong reciprocation asserts that many humans are willing to do it, even if they know that they will not meet the punished (and possibly reformed) wrong-doer ever again. Strong reciprocators contribute to the public good, and punish those who don't. There exist several attempts to explain this trait (e.g. Gintis 2000, Fehr and Fischbacher 2003) of which at least one, incidentally, is based on reputation (Sigmund et al. 2001). In the context of indirect reciprocation, we can view discrimination as a form of punishment: low-scorers are deprived of help. If players assess each other according to their standing, the punishment is not costly for the punisher. But if they register only whether the other defected or not, without distinguishing between justified and non-justified defections, then punishment is costly. In view of the fact that many humans are ready to engage in costly punishment in a great variety of contexts (see e.g. Fehr and Gächter 2002), it cannot altogether be excluded that this factor also plays a role in indirect reciprocation. As we shall see in the last section, experiments support this view (Milinski et al. 2001).

On theoretical grounds, it is therefore not obvious how individuals update the scores of their co-players. In fact, this standard of moral judgement, which eventually leads to a social norm, can also be subject to evolution.

In the following investigation we shall assume that individuals engaged in the indirect reciprocation game keep track of the scores in their community, and then decide, when in the role of the donor, whether to give help or not, depending on the recipient's score, and possibly on their own. Needless to say, one can envisage many other strategies, taking into account the accumulated payoffs for donor and recipient, the prevalence of cooperation within the community, the outcome of the last round as a recipient etc. We shall not consider these possibilities in the following models, but start by describing the recent results obtained in two papers, one by Ohtsuki and Iwasa (2004), the other by Brandt and Sigmund (2004), which both, independently, adress the issue of the evolution of updating mechanisms for the indirect reciprocity game. This can be viewed as investigating simple mechanisms for local information processing. But it has farther-ranging implications for the evolution of social norms, and hence of moral judgements. When is a defection justified, or not? When is a player good, or bad? Let us first consider this question in a very limited context, when the score can only take two values.

3.5 Binary models

We shall assume that every strategy consists of two modules, an assessment module and an action module. The assessment module comes into play when individuals observe interactions between two players. The image of the player acting as potential donor is possibly changed. The image of the recipient, who is the passive part in the interaction, remains unchanged. The action module prescribes whether a player in the position of a potential donor provides help or not, based on the information obtained through the player's assessment module.

Starting with the assessment module, we shall for simplicity assume that individual A's score of individual B depends only on how B behaved when last observed by A as a potential donor, i.e. whether B gave or refused help to some third party C. Thus A has a very limited memory, and the score of B can only take two values, *good* and *bad*. (In this context, we note that Dellarocas (2003) found that binary feedback mechanisms publishing only the single most recent rating obtained by an online seller are just as efficient as mechanisms publishing the sellers total feedback history). We shall assume that all players are born *good*. In every interaction observed by A, there are two possible outcomes (B can give help or not), two possible score values for B and two for C. Thus there are eight possible types of interaction, and hence, depending on whether they find A's approval or not, $2^8 = 256$ different value systems.

As intuitively appealing examples of such assessment modules, let us consider three of these value systems, or 'morals'. We shall say that they are based on SCORING, STANDING and JUDGING, respectively (these terms are not completely felicitous, but the names of the first two, at least, are fixed by common use). These morals differ on which of the observed interactions incur reprobation, i.e. count as *bad*. Someone using the SCORING assessment system will always frown upon any potential donor who refuses to help a potential recipient, irrespective of the latter s image. Someone using the STANDING assessment system will condemn those who refuse to help a recipient with a *good* score, but will condone those who refuse to help a recipient with a *bad* score. Those using the JUDGING assessment system will, in addition, extend their reprobation to players who help a co-player with a *bad* score.

Thus these three value systems are of different strictness towards wrongdoers. Roughly speaking, someone who refuses to help is always bad in the eyes of a SCORING assessor. Only those who fail to give to a *good* player are bad in the eyes of a STANDING assessor. Someone who fails to give to a *good* player, but also someone who gives help to a *bad* player is bad in the eyes of a JUDGING assessor (see Table 3.1).

Turning to the action module, we shall assume that a player's decision on whether to help or not is based entirely on the scores of the two players involved. Since there are four situations (donor and recipient can each be *good* or *bad*), there are $2^4 = 16$ possible decision rules. Four intuitively appealing examples would be CO, SELF, AND and OR. CO is uniquely affected by the score of the potential recipient, and gives if and only if that score is *good*. SELF worries exclusively about the own score, and gives if and only if this

Table 3.1. The assessment module specifies which image to assign to the potential donor of an observed interaction ('good → bad' means 'a good player helps a bad player', 'bad ↛ good' means 'a bad players refuses to help a good player', etc)

Assessment Modules

situation/strategy	SCORING	STANDING	JUDGING
good → good	good	good	good
good → bad	good	good	bad
bad → good	good	good	good
bad → bad	good	good	bad
good ↛ good	bad	bad	bad
good ↛ bad	bad	good	good
bad ↛ good	bad	bad	bad
bad ↛ bad	bad	bad	bad

Table 3.2. The action module prescribes whether to help or not given the own image, and the image of the potential recipient ('bad $\xrightarrow{?}$ good' prescribes whether a bad player should help when faced with a good co-player, etc.)

Action modules

situation/strategy	SELF	CO	AND	OR	AllC	AllD
good $\xrightarrow{?}$ good	no	yes	no	yes	yes	no
good $\xrightarrow{?}$ bad	no	no	no	no	yes	no
bad $\xrightarrow{?}$ good	yes	yes	yes	yes	yes	no
bad $\xrightarrow{?}$ bad	yes	no	no	yes	yes	no

score is *bad*. AND gives aid if the recipient's score is *good* and the own score *bad*, and OR gives aid if the recipient's score is *good* or the own score *bad*. Of course the 16 decision rules also include the two unconditional rules, always to give, and never to give, ALLC and ALLD, which do not rely on scores at all. (see Table 3.2).

A strategy in this model for indirect reciprocity is determined by a specific combination of action and assessment module. This yields altogether $2^4 \times 2^8 = 2^{12} = 4096$ different strategies.

3.6 The leading eight

Ohtsuki and Iwasa (2004) have investigated the evolutionary stability of these strategies. Thus they looked for strategies with the property that a population whose members all use this strategy cannot be invaded by a small minority using another strategy. Ohtsuki and Iwasa assumed that players were subject to errors, by implementing an unintended move (with a probability μ) or by assigning an incorrect score to a player (with probability ν). Depending on the values of μ, ν, b and c, they found various evolutionarily stable strategies (ESS), including of course ALLD. Most remarkably, they singled out eight strategies (called 'the leading eight') which are robust against errors and lead to cooperation even if b is only slightly larger than c (the ratio must exceed 1 by a factor proportional to the error probabilities).

Only the CO and the OR action module occur among the leading eight. Such players always give help to a *good* player, and defect (when *good*) against a *bad* player. The assessment module of the leading eight is consistent with this prescription: they all assess players as *good* or *bad* if they give (resp. withhold) help to a *good* player, irrespective of their own score, and they all allow *good* players to refuse help to *bad* players without losing their reputation. Interestingly, all other actions towards a *bad* player are possible, i.e. whether a *good* player gives help to a *bad* player, or a *bad* player gives (or refuses) help to a *bad* player. These are just the eight alternatives making up the leading eight. If the assessment module requires a *bad* player to give to a *bad* player, the corresponding action module is OR; in all other cases it is CO. We note that strategies with the STANDING and the JUDGING assessment module can belong to the leading eight, but not those with the SCORING module.

It seems obvious that in an ESS leading to cooperation, assessment rules and action rules should correspond. This requirement does not hold for CO-SCORING, for instance, where *good* players have to refrain from helping *bad* players although this makes them lose their *good* score. Interestingly, there is one exception to this requirement, among the leading eight: for the last two strategies displayed on Table 3.3, *bad* players meeting *bad* co-players cannot redress their score one way or the other. However, in a homogenous population playing this strategy, encounters between two *bad* players are exceedingly rare.

Ohtsuki and Iwasa obtained their analytical results under the assumption that players experience infinitely many interactions during their life-time (an approximation which implies that the population is very large). Furthermore, they demand from their ESS strategies only that they are able to repel invasions by strategies with the same assessment module. They also assume that a player's score is the same in the eyes of all co-players. This last assumption is justifed by the so-called 'indirect observation model', which postulates that an interaction between A and B, say, is observed by one player only, for instance C, and that all other members of the population adopt C's assessment. A similar model is used in Panchanathan and Boyd (2003). Other authors, for instance Nowak and Sigmund (1998), Lotem and Fishman (1999) or Leimar and Hammerstein (2001), adopt a 'direct observation model' where all players keep their own, private score of their co-players. Ultimately, it would seem that the evolution of assessment modules will have to be addressed in this context. It is argued that thanks to language, all members of a popu-

Table 3.3. The leading eight ESS strategies, specified by as assessment module (first 8 rules) and an action module (last 4 rules), obtain highest payoffs among all ESS pairs, and keep their evolutionary stability even for benefit-to-cost ratios close to one. Strategy 1 corresponds to OR-STANDING (Contrite Tit For Tat, or CTFT, in Panchanathan and Boyd, 2003), strategy 8 corresponds to CO-JUDGING. Note that neither CO-STANDING, the RDISC strategy from Panchanathan and Boyd, 2003, nor any SCORING strategy occurs in the list

The leading eight

situation/strategy	1	2	3	4	5	6	7	8
good → good	good	good	good	good	good	good	good	good
good → bad	good	bad	good	good	bad	bad	good	bad
bad → good	good	good	good	good	good	good	good	good
bad → bad	good	good	good	bad	good	bad	bad	bad
good $\not\to$ good	bad	bad	bad	bad	bad	bad	bad	bad
good $\not\to$ bad	good	good	good	good	good	good	good	good
bad $\not\to$ good	bad	bad	bad	bad	bad	bad	bad	bad
bad $\not\to$ bad	bad	bad	good	good	good	good	bad	bad
good $\overset{?}{\to}$ good	yes	yes	yes	yes	yes	yes	yes	yes
good $\overset{?}{\to}$ bad	no	no	no	no	no	no	no	no
bad $\overset{?}{\to}$ good	yes	yes	yes	yes	yes	yes	yes	yes
bad $\overset{?}{\to}$ bad	yes	yes	no	no	no	no	no	no

lation should agree on their scores, and it may well be indeed that gossip is powerful enough to furnish all individuals with information about all past interactions. But it is common-day experience that even if two people witness the same interaction directly, they can differ in their assessment of that interaction. This strongly argues for private scores, and has strong implications: as Ohtsuki and Iwasa stressed, CO-STANDING is not an ESS in the direct observation model, but can be invaded, if errors in perception occur, by the undiscriminating ALLC.

3.7 Replicator dynamics

Another way to approach analytically the evolution of indirect reciprocity is via replicator dynamics. For this, one clearly has to drastically reduce the number of strategies involved. Typically, one considers only three: ALLC, ALLD and a discriminating strategy. Indeed, the main problem for the emergence of discriminating cooperation is that it is threatened by strategies which do not punish defection, and eventually undermine the stability of the helping behavior.

The discriminating strategy usually investigated in this context is CO-SCORING. Let us assume that each player has two interactions per round, one as a donor and one as a recipient, against two different, randomly chosen co-players. (Assuming one interaction only, with equal probability as donor or recipient, changes the expressions but not the conclusions). We denote the frequency of the indiscriminate altruists, i.e. the ALLC-players, with x, that of defectors, i.e. the ALLD-players, with y, and the frequency of the discriminate altruists, i.e. the CO-SCORING players, with $z = 1 - x - y$. To begin with, we assume that in the first round, discriminators consider their co-players as *good*. With $P_x(n)$, $P_y(n)$ and $P_z(n)$ we denote the expected payoff in the n-th round for ALLC, ALLD and CO-SCORING, respectively. It is easy to see that

$$P_x(1) = -c + b(x+z),$$
$$P_y(1) = b(x+z),$$

and

$$P_z(1) = -c + b(x+z).$$

In the n-th round (with $n > 1$) it is

$$P_x(n) = -c + b(x+z),$$
$$P_y(n) = bx$$

and

$$P_z(n) = -cg_n + b(x + zg_{n-1})$$

where g_n denotes the frequency of *good* players at the start of round n (with $g_1 = 1$) and g_{n-1}, therefore, is the probability that the discriminator has met a *good* player in the previous round. Clearly $g_n = x + zg_{n-1}$ for $n = 2, 3, \ldots$ (the *good* players consist of the ALLC players and those discriminators who have met players with a *good* score in the previous round). Hence

$$P_z(n) = (b-c)g_n$$

and by induction

$$g_n = \frac{x}{x+y} + z^{n-1}\frac{y}{x+y}.$$

In the limiting case $n \to +\infty$ this yields

$$P_z = (b-c)\frac{x}{1-z}.$$

If there is only one round per generation, then defectors win, obviously. This need no longer the case if there are N rounds, with $N > 1$. The total payoffs $\hat{P}_i := P_i(1) + \cdots + P_i(N)$ are given by

$$\hat{P}_x = N[-c + b(x+z)], \quad \hat{P}_y = Nbx + bz,$$

and
$$\hat{P}_z = N(b-c) + y[-b + \frac{b-c}{1-z}(1+z+\ldots+z^{N-1} - N)].$$

Let us now assume that the frequencies of the three strategies evolve under the action of selection, with growth rates given by the difference between their payoff $\hat{P}_i$ and the average $\hat{P} = x\hat{P}_x + y\hat{P}_y + z\hat{P}_z$. This yields the replicator equation $\dot{x} = x(\hat{P}_x - \hat{P})$, $\dot{y} = y(\hat{P}_y - \hat{P})$ and $\dot{z} = z(\hat{P}_z - \hat{P})$ on the unit simplex S_3 spanned by the three unit vectors e_x, e_y and e_z of the standard base.

In there are exactly N rounds in the game, this equation has no fixed point with $x > 0, y > 0$ and $z > 0$, hence the three types cannot co-exist in the long run. The fixed points are: the defectors corner e_y with $y = 1$; the point F_{yz} with $x = 0$ and $z + \ldots + z^{N-1} = c/(b-c)$; and all the points on the edge $e_x e_z$. Hence in the absence of defectors, all mixtures of discriminating and indiscriminating altruists are fixed points.

The overall dynamics can be most easily described in the case $N = 2$ (see Fig. 3.1).

The parallel to the edge $e_x e_y$ through F_{yz} is invariant. It consists of an orbit with ω-limit F_{yz} and α-limit F_{xz}. This orbit l acts as a separatrix. All orbits on one side of l converge to e_y. This means that if there are too few discriminating altruists, i.e. if $z < c/(b-c)$, then defectors take over. On the other side of l, all orbits converge to the edge $e_x e_y$. In this case, the defectors are eliminated, and a mixture of altruists gets established.

This leads to an interesting behaviour. Suppose that the society consists entirely of altruists. Depending on the frequency z of discriminators, the state is given by a point on the fixed point edge $e_x e_z$. We may expect that random drift makes the state fluctuate along this edge and that from time to time, mutation introduces a small quantity y of defectors. What happens then? If

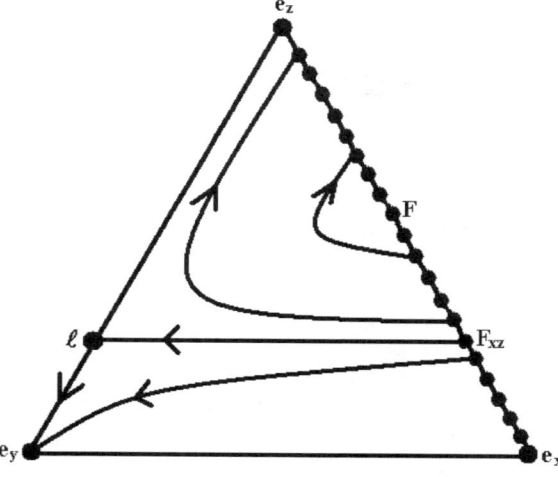

Fig. 3.1. Replicator dynamics when the number of rounds is constant. In the absence of errors, any mixture of AllC and CO-SCORING is a fixed point

the state is between F_{xz} and e_x, the defectors will take over. If the state is between e_z and F, the state with $z = 2c/b$, they will immediately be selected against, and promptly vanish. But if a minority of defectors invades while the state is between F and F_{xz}, then defectors thrive at first on the indiscriminating altruists and increase in frequency. But thereby, they deplete their resource, the indiscriminating altruists. After some time, the discriminating altruists take over and eliminate the defectors. The population returns to the edge $e_x e_z$, but now somewhere between e_x and F, where the ratio of discriminating to indiscriminating altruists is so large that defectors can no longer invade. The defectors have experienced a Pyrrhic victory. They can only take over if their invasion attempt starts when the state is between F_{xz} and e_x. For this, the fluctuations have to cross the gap between F and F_{xz}. This takes some time. If defectors try too often to invade, they will never succeed.

In the limiting case that the number of rounds N is infinite, we obtain for the average payoffs P_i per round, that

$$P_x - P_y = bz - c$$

and

$$P_z - P_y = \frac{x}{1-z}(P_x - P_y).$$

In the interior of S_3, the fixed points form a line $z = c/b$ parallel to the edge $e_x e_y$. We denote this line by l (it is just the limit of the separatrix l in the previous paragraph, for $N \to +\infty$). The edges with $x = 0$ and $y = 0$ consist of fixed points. In the interior of S_3 all orbits are parallel to l. Those with $z < c/b$ converge to the left (the discriminating altruists vanish), those with $z > c/b$ to the right (the undiscriminate altruists vanish) (see Fig. 3.2).

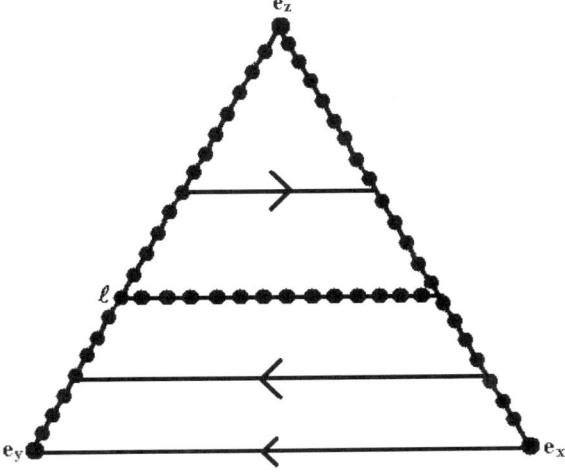

Fig. 3.2. Replicator dynamics in the limiting case of infinitely many rounds, and no errors. In addition to the fixed point edges, we obtain a line of fixed points in the interior of the simplex

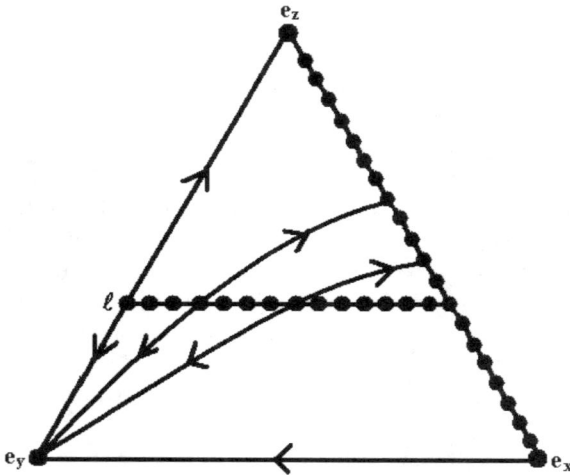

Fig. 3.3. Replicator dynamics when the number of rounds follows a geometric distribution and no errors occur

If there is a fixed probability $w < 1$ for a further round (see Nowak and Sigmund, 1998b), we obtain for the total payoff values:

$$P_x - P_y = \frac{wbz - c}{1 - w}$$

and

$$P_z - P_y = \frac{1 - w + wx}{1 - wz}(P_x - P_y)$$

and the fixed points form the line l defined by $z = c/wb$, as well as the $\mathbf{e}_x\mathbf{e}_z$-edge. In the interior of S_3 the orbits are on the curves with $z = ax^{1-w}$ (see Fig. 3.3).

Above l the orbits converge to the fixed point edge with $y = 0$, below l to the vertex with $y = 1$. The state will drift along the fixed point lines until a mutation sends it to the region below l, where the defectors win.

It is clear that such a degenerate behaviour is rather sensible to perturbations. Let us assume that errors in implementation can occur. For simplicity, we consider only errors turning an intended cooperation into a defection with a certain probability $1 - r$. Equivalently we may assume, following Lotem et al. (1999), that $1 - r$ is the probability that an individual is actually unable to perform the intended act of giving help (this incapacity may be due, for instance, to a lack of resources or an injury). Such an incapacity is highly likely: as Fishman (2004) wrote, individuals who are always able to help do not need help from others... In practice, one donates help when the costs are small, in order to secure reciprocity in the hour of need. 'The defectors' payoff in the first round is $P_y(1) = rb(x + z)$, and in all further rounds it is $P_y(n) = rbx$. In the n-th round ($n > 1$) we obtain $P_x(n) = -rc + br^2 z + P_y(n)$, and $P_z(n) = -rcg_n + rbzrg_{n-1} = r(b-c)g_n - br^2 x + P_y(n)$, where g_n, the frequency of players with a *good* image at the start of the n-th round, satisfies

$g_n = r(x + zg_{n-1})$ and is given by

$$g_n = \frac{rx}{1-rz} + (rz)^{n-1}\frac{1-rx-rz}{1-rz}$$

(clearly $g_1 = 1$ and $P_x(1) = P_z(1) = -rc + P_y(1)$). These expressions have been obtained by Panchanathan and Boyd (2003) and by Fishman (2004). In the limiting case of infinitely many rounds,

$$P_z - P_y = \frac{rx}{1-rz}(P_x - P_y).$$

Once more, we obtain a line l of equilibria in the interior of S_3, given by $z = c/br$. This line intersects the edge $e_x e_z$ (where all players are altruists) at the point $\boldsymbol{F}_{xz}$. This time, the edge does not consist of fixed points: in fact $\boldsymbol{F}_{xz}$ is the only equilibrium mixture of discriminating and indiscriminating altruists, and it is stable within the edge $e_x e_z$ of altruists. Indeed, if almost all altruists are indiscriminating, then the unintended defections which cause discriminating altruists to refuse help in the next round will allow them to obtain a higher payoff than ALLC-players without being taken to account too frequently; whereas if most altruists are discriminating, most refusals to help will be severely punished. If we consider an arbitrary mixture of defectors and altruists, the orbit will either converge to e_y or it will first converge to the line l, drift along this line and then, if a random shock introduces some more defectors while the frequency of undiscriminating altruists is sufficiently low, to e_y. In any case the evolution will ultimately lead to the fixation of defectors (see Fig. 3.4).

Panchanathan and Boyd (2003) have noticed that the same happens if the number of rounds is not infinite, but a random variable with a geometric distribution given by a parameter w (a constant probability for a further round).

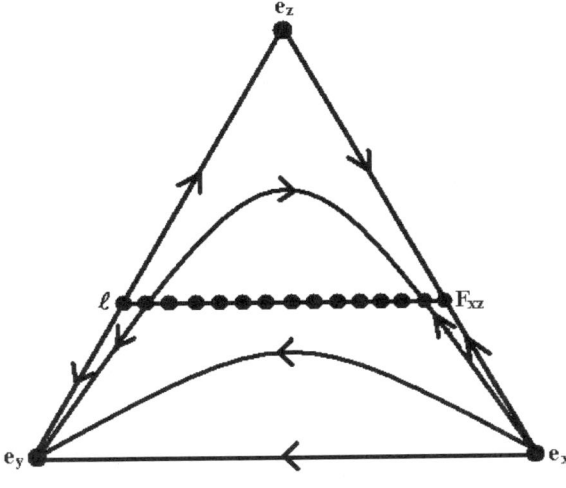

Fig. 3.4. Replicator dynamics if individuals make errors in implementation, and the number of rounds follows a geometric distribution. In the long run, AllD is established. A similar dynamics holds for the asynchronous entry case, for all probability distributions of rounds

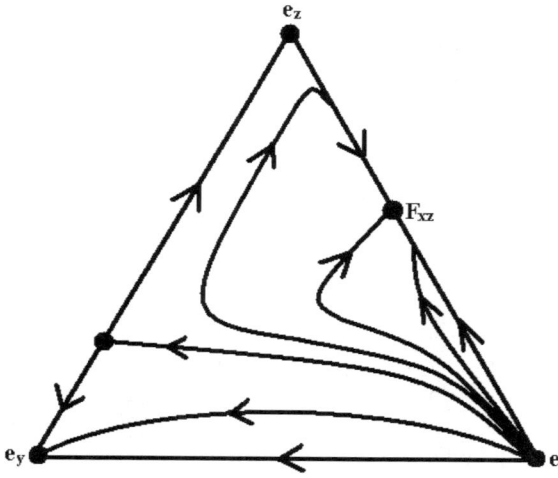

Fig. 3.5. Replicator dynamics when individuals make errors in implementation and the number of rounds is constant. A bistable outcome results. The same holds if the rounds are Poisson distributed, or in the asynchronous entry case when each player's social network grows with time

In contrast, they found that if the discriminating strategy is OR-STANDING or CO-STANDING, then the monomorphic state with all players discriminating is stable. They also found that OR-STANDING is slightly superior to CO-STANDING (which does not belong to the 'leading eight', incidentally). Panchanathan and Boyd concluded that 'when errors are added, indirect reciprocity cannot be based on an image-scoring strategy'. And indeed they have pointed out an important vulnerability of the CO-SCORING strategy. Nevertheless, the verdict seems to depend on the modelling assumptions.

Indeed, as shown by Fishman (2004), if one assumes that the number N of rounds is constant, then the equilibrium $\boldsymbol{F}_{xz}$ is transversally stable, i.e. it cannot be invaded by defectors if $c/b < 1 - 1/N$, and if r is sufficiently large: for this, one has only to check that the payoff values $P_x = P_z$ at $\boldsymbol{F}_{xz}$ exceed P_y. Hence cooperation can be stably sustained. Brandt and Sigmund (2004) showed that the same holds if the number of rounds is a random variable with a Poisson distribution with parameter λ, provided $b > 2c$ and λ is sufficiently large. In both cases, the model leads to a bistable dynamics (see Fig. 3.5).

Depending on the initial condition, either defectors take over, or the population converges to a stable mixture of discriminating and undiscriminating altruists, and hence to a cooperative regime.

Fishman stressed, therefore, that involuntary defection (caused by errors, or by incapacitation) stabilises indirect reciprocity. He states: 'Indirect reciprocity, at least in the current case, is stable only among imperfect individuals.' In Lotem et al. (1999), Lotem et al. (2002), and Sherratt and Roberts (2001), this inability of giving help, due to lack of quality, is further analysed: helping behaviour is used as a way of signalling high quality (see also Zahavi, 1995).

Ohtsuki (2004) studied adaptive dynamics for stochastic strategies of the CO-SCORING type. His strategies are given by triples (p_0, p_1, p_2) where p_0

is the probability to help a player in the absence of information about his score (an event whose likelihood is q), whereas p_1 and p_2 are the probabilities to help an individual with *good* resp. *bad* score. In his analysis of monomorphic populations, Ohtsuki finds that there exist two regions, one in which all p_i-values increase and one in which all decrease. As in the case of direct reciprocation (see Nowak and Sigmund, 1990), the discriminating strategies (with $p_1 = 1$ and $p_2 = 0$) act not as end-points but rather as pivots of the evolution: in their neighborhood all p_i-values increase but the degree of discrimination $p_1 - p_2$ decreases so that eventually a continuum of equilibria is approached. Once there, mutations can send the population towards defection. This instability is even more pronounced if errors in perception or implementation are included.

3.8 Asynchronous entry

So far, we have assumed that the whole population lives according to the same schedule: all players engage together in the first round (once as donor and once as recipient), then all in the second round etc... This can indeed model what happens with a group of persons volunteering for an experimental game. But for real-life interactions, it may seem more appropriate to model a population with generations blending into each other. Occasionally, a new player is born, and will from time to time play a round of the game. In contrast to the previous model, different players will usually experience a different number of rounds: these rounds are no longer synchronised. Let us denote by g the frequency of individuals with *good* reputation in the population. If the population is sufficiently large, and stationary, then g will not be affected by the birth of a new individual, or its age. Let us assume, to begin with, that newcomers are considered as *good*. After the first round, an ALLC player will have a *good* reputation with probability r, an ALLD player with probability 0 and a discriminator with probability rg. Hence $g = rx + rzg$ and thus

$$g = \frac{rx}{1 - rz}.$$

The payoff for an ALLC player is $-cr + br(x + z)$ in the first round and $-cr + br(x + zr)$ in all following rounds. For an ALLD player it is $br(x + z)$ in the first round and brx in all following rounds. For a discriminator the payoff is $-cgr + br(x + z)$ in the first round and $-cgr + br(x + gzr)$ in all following rounds. Thus $P_z - P_y = g(P_x - P_y)$.

In the limiting case of infinitely many rounds we see that

$$P_z - P_x = r(1-g)(c - brz)$$

which yields again a line l of fixed points satisfying $z = c/br$. The phase portrait looks like that of Fig. 3.4 (if $r > c/b$), and defectors will always win in the end. This holds also if the number of rounds is any random variable with expectation value E, except that the z-value of l has to be multiplied with a certain factor. Indeed, since $P_x = P_z$ holds if and only if $P_x = P_y$, it follows that $P_x = P_z$ always defines a line of fixed points.

A similar result is obtained in a model where discriminators know their co-player's reputation (i.e. their behaviour in the last round) only with a certain probability q, and assume that it is *good* if they have no information We note in this context that Panchanathan and Boyd (2003) have shown that for sufficiently large q and b/c, it is selectively advantageous to be trustful in this sense. For ALLC players, the payoff is $-cr + br(x + z)$ in the first round and $-cr + br[x + (1-q)z + qzr]$ in all subsequent rounds. For ALLD players, it is $br(x + z)$ in the first and $br[x + (1-q)z]$ in all subsequent rounds. For discriminators, it is $-cr(1-q) - crqg + br(x+z)$ in the first round and $-cr(1-q) - crqg + brx + br(1-q)z + brqzr[(1-q) + qg]$ in all subsequent rounds. ALLC players and discriminators have the same payoff iff $z = c/br$, in the limiting case of infinitely many rounds. Since

$$P_x - P_z = rq(1-g)(P_x - P_y)$$

holds in every round, there exists, for sufficiently large r, a stable equilibrium mixture of discriminating and undiscriminating altruists, but defectors can invade through random drift. The same holds also for other scoring strategies, as for instance for OR-SCORING; it also holds if we assume that a discriminator who does not know the recipient's score defects. Thus we see that the argument of Panchanathan and Boyd (2003) is even more robust for the asynchronous entry case than for the case of synchronised rounds: it holds whenever the probability that a discriminator gives help is the same from one round to the next.

But assume now that q_n, the probability that a discriminator engaged in round n knows the score of the co-player, is increasing in n. This assumption is plausible: with time, a player's social network grows, and therefore also the player's probability to have information about the recipient. Of course, if the population has reached a steady state, then the average probability that a randomly chosen player knows a co-player's score is just the mean value of the q_n, i.e. some constant q. If we assume, as before, that discriminators are trustful, in the sense that they provide help if they do not know the co-player's score, then we obtain as payoffs in the n-th round:

$$P_x(n) = -cr + brx + br(1-q)z + br^2qz$$
$$P_y(n) = brx + br(1-q)z$$

and

$$P_z(n) = -cr[(1-q_n) + q_n g] + brx + br(1-q)z \\ + brqzr[(1-q_{n-1}) + q_{n-1}g].$$

Thus
$$P_x(n) - P_y(n) = -cr + br^2 qz$$
and
$$P_z(n) - P_y(n) = P_x(n) - P_y(n) + r(1-g)[cq_n - brzqq_{n-1}].$$
Clearly
$$P_x(n) - P_z(n) = r(1-g)(-cq_n + zbrqq_{n-1}).$$
Let $w(n)$ be the probability that a randomly chosen donor is in round n, then
$$q = \sum w(n)q_n > \hat{q} := \sum w(n)q_{n-1}.$$

In Brandt and Sigmund (2005) it is shown that with $z_{cr} = \frac{c}{br\hat{q}}$, there exists a mixture of discriminating and indiscriminating altruists $\boldsymbol{F}_{xz} = (1 - z_{cr}, 0, z_{cr})$ which is a fixed point. For sufficiently small $w(1)$ (i.e. a sufficiently large likelihood of having more than one round)
$$P_x(z_{cr}) > P_y(z_{cr}).$$

Hence $\boldsymbol{F}_{xz}$ cannot be invaded by the defectors. The resulting replicator equation is bistable: one attractor consists of defectors only, the other of a mixture of discriminating and undiscriminating altruists.

If $q_n < q_{n-1}$ this would not be valid: except if we assume that discriminators who do not know the recipient's score, instead of helping, i.e. according the benefit of doubt, prefer to refuse help, i.e. to act distrustfully. To resume, we see that if either players are trusting and have a growing social net, or if they are distrustful and have a shrinking net of acquaintances, a stable mixture of discriminating and indiscriminating altruists can be supported by the SCORING assessment module.

3.9 Numerical simulations

It seems hard to derive analytical expressions for the payoff values if several discriminating strategies are present, and errors in perception and implementation, limited observability etc are taken into account. Thus while it is easy to compute the payoff expressions for mixtures of CO-SCORING with ALLC and ALLD, merely adding OR-SCORING or CO-STANDING to the cast greatly complicates things. Often, pairs of discriminating strategies perform equally well against each other, so that their frequencies drift randomly around: but the success of other strategies at invading them depends on their frequencies, etc. One is often reduced to numerical simulations to investigate such polymorphic states.

In Nowak and Sigmund (1998a,b), well-mixed populations are considered, consisting of some 100 individuals each engaged in some five or ten interactions, sometimes as as a donor, and sometimes as a recipient. But in order to avoid spurious effects of random drift, it is convenient to adopt, following Leimar and Hammerstein (2001), a population structure conveying a more realistic image of prehistoric mankind, and consider some 100 tribes, for instance, with 100 players each, with some modest gene flow between the tribes. We shall start by describing the extensive statistical investigations of Brandt and Sigmund (2004), based on such a population structure, and the assumption of a binary score.

Let us consider the case of separate generations. During one generation, there will be 1000 games within each tribe, so that on average each player is engaged in 10 rounds (a larger number does not significantly change the outcome). Each individual keeps a private score of all tribe-members. We normalise payoffs by setting $c = 1$, so that b is now the cost-to-benefit ratio. At the end of each generation, each tribe forms a new generation of 100 individuals: with probability p the new individual will be 'locally derived' and inherit a strategy from a member of the tribe, and with probability $1-p$, the new individual will inherit a strategy from some member at large, in each case with a probability which is proportional to that member's total payoff. In order to avoid transitional effects, we present averages over 1000 generations, after an initial phase of 9000 generations. (Usually, a stable composition is reached within 100 generations). In Brandt (2004) one can find an online approach to such numerical simulations which allows the visitors of that site a great deal of experimentation.

Let us first ask which strategies are best at invading a population of defectors, when introduced as a minority of, for instance, 10 percent. It turns out that in the absence of errors, STANDING and JUDGING, together with the CO and the OR module, do best and lead to cooperation whenever $b \geq 4.5$, whereas SCORING requires considerably higher b-values. In the presence of errors, this is attenuated: if, for instance, ALLC, ALLD and a single discriminating strategy are initially equally frequent, then CO-STANDING and OR-STANDING eliminate defectors whenever $b > 3.5$, whereas CO-JUDGING and CO-SCORING require $b > 4.5$, and OR-JUDGING and OR-SCORING even $b > 6.5$.

If a given assessment module is held fixed and several action-modules start at similar frequencies, then cooperation dominates for STANDING and for SCORING as soon as $b > 4$, usually with the CO or the OR module (together with a substantial ALLC population). Less cooperative action modules, as for instance SELF or AND, are rapidly eliminated.

There is a strong propensity for cooperation based on polymorphisms. Let us, for instance, start with a population where the three assessment modules SCORING, STANDING and JUDGING as well as the action modules AND, OR, CO and SELF, together with the indiscriminate strategies ALL C and ALL D are present in equal frequencies. Even if only every second interaction is observed, a cooperative outcome is usually achieved as soon as $b > 2.5$, and CO-SCORING, OR-SCORING, CO-STANDING and OR-STANDING prevail at nearly equal frequencies. JUDGING is greatly penalised by the lack of reliable information. On the other hand, if all interactions are observed and only errors in implementation occur, then CO-JUDGING and OR-JUDGING dominate, eliminating ALLC players and establishing a very stable cooperative regime. If errors in perception occur, then JUDGING is completely eliminated, and SCORING and STANDING perform on a similar level. This also holds if errors in implementation or limited observability are taken into account.

In a recent and as yet unpublished paper, Takahashi and Mashima (2004) have shown that STANDING is highly vulnerable to errors in perception, if one does not consider a subdivided population linked by migration, as in Leimar and Hammerstein, but a single well-mixed tribe. On the other hand, they emphasised the success of a strategy which had not been considered before, and in particular is not a member of the 'leading eight'. Its action module is CO, and its assessment module ascribes a *bad* score, not only to those refusing help to a *good* player, but to all those who interacted with a *bad* player (irrespective of whether they provided help or not). Players who have met a *bad* player are *bad* and remain so until they are able to redeem themselves by giving to a *good* player. According to Takahashi and Mashima, it remains still to be checked whether such intriguing strategies can get established in more polymorphic populations.

3.10 Spatial indirect reciprocation

In a variant of evolutionary games, spatially distibuted populations are considered, with each individual interacting only with the closest neighbors and updating by switching to the strategy of a random neighbor with a probability proportional to the payoff difference. Let us assume, for instance, that the players sit on an $N \times N$-lattice, with the usual identification of opposite borders, and that the neighborhood of site (i,j) consists of the 8 sites whose coordinates differ by at most one unit (a Moore neighborhood). Since the score depends on how many games have already been played, it is important to introduce no systematic bias in the ordering of the games. A simple approach is to arrange all individuals in a random sequence and let the interactions take place in that order, with this individual as recipient, and one of the neighbors (randomly chosen) as potential donor. Individuals cannot receive help more than once per round, but they may be asked more than once to help a co-player. Not surprisingly, the spatial games lead to the evolution of cooperation for even smaller b/c-values than in the well-mixed case (see Fig. 3.6 and, for interactive experimentation, Brandt 2004). In these simulations, a small mutation probability and a probability of not being able to cooperate, due to lack of resources for example, is included. Moreover, discriminators are tempted to defect instead of helping with a small temptation rate. Every generation consists of 5 rounds played as described. Then, in the spatial case, sites are updated by comparing their payoff with that of a ran-

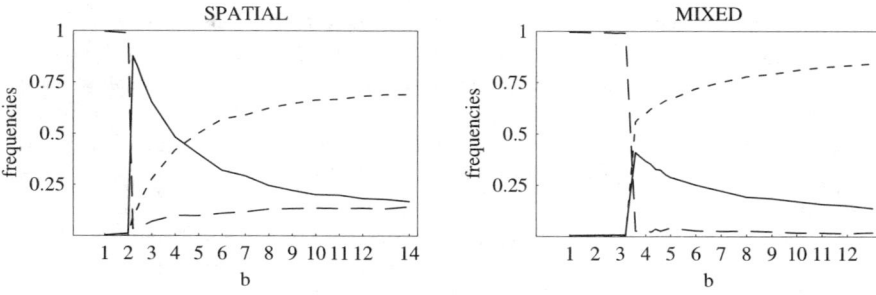

Fig. 3.6. In both graphs, long-term frequencies for a population initialized randomly with strategies AllC (short dashes), AllD (long dashes), and CO-SCORING (solid line) are shown, a mutation rate 0.001, an error rate of 0.05, and a temptation rate for discriminators to defect of 0.05 are included. Five rounds per generation are played. Spatial indirect reciprocity, where individuals are confined to the sites of a square lattice and interact only with their neighbors, promotes cooperation for smaller benefits (with $c = 1$) than in the well-mixed case. In the spatial case, however, defectors can survive more easily within clusters of AllC players, and subsist at frequencies of around 15%

domly chosen neighboring site (a randomly chosen site of the full lattice, in another variant) and switching to the strategy at that site with a probability proportional to the payoff difference, if the own payoff is lower. In the spatial case, when updating occurs only between neighbors, cooperation dominates for $b/c \geq 2$, whereas if the population is well-mixed, it takes $b/c \geq 3.5$ to suppress defectors to a small minority.

3.11 Full score

The original numerical simulations of Nowak and Sigmund (1998a) considered the case, not of a binary score, but of a full score ranging through all integer values. This means that if, on average, individuals experience only five rounds as a donor, their score cannot exceed ± 5. This score range seems much more natural than the restriction to a binary score. In fact, binary scores were only introduced as a crude simplification to allow for analytical results.

With a full score, one can again consider the same assessment modules as before, and in particular SCORING or STANDING. One can also consider different action modules, but their number vastly increases. For instance, in the OR-family, we would find all strategies of the type $(k \vee h)$, meaning 'help if the recipient's score exceeds k or if your own score is below h'. It seems intuitively clear that the main disadvantage of SCORING, namely that punishing is costly, is greatly reduced. Indeed, players with a high reputation for helping will be able to refrain occasionally from helping a low-scorer without threatening their own score, which will be reduced by one unit but remain in the high range. The numerical simulations of Nowak and Sigmund (1998a) were confirmed by Leimar and Hammerstein (2001), who found, however, that a modest gene-flow between groups reduced the success of SCORING. Thus while, for $b/c = 4$, AND-SCORING produces on average 40 percent of cooperation in isolated groups without migration, it does much less well if mixing occurs between the groups. We note that the poor showing of the AND-module is also reflected in the simulations with a binary module. Such strategies are not cooperative in the sense that they do not always lead to help-giving in a monomorphic population.

Leimar and Hammerstein also reported an interesting robustness of the STANDING module against errors of perception, adding that the issue was not fully resolved yet. Indeed, a systematic investigation of the different assessment modules for the full-score case is lacking so far, due in part to analytical difficulties, and in part to the fact that the proliferation of strategies for each assessment module often leads to neutral polymorphisms which are dominated by random drift rather than a clear-cut selective force. It seems safe to predict that the costs of complexity, the prevalence of phenotypic defectors (i. e. players unable to give help even if they want to) and the issue of public vs private scores will become essential topics for these investigations.

In an interesting approach, Mohtashemi and Mui (2003) have performed agent-based simulations based on the SCORING module, using players with growing networks of acquaintances (in every round, the donors and their acquaintance are added to the acquaintance of the recipients). They found that this greatly promotes the emergence of cooperation, a result which agrees well with our analysis of the replicator dynamics in the asynchronous entry case.

3.12 Experimental games

Wedekind and Milinski (2000) set up experiments with 79 undergraduate students, who were divided into eight groups. All players were provided with a starting account, and were repeatedly offered the possibility to give 4 Swiss Francs to another person of the same group, at a cost of 1 (or, in some groups, 2) Swiss Franc to themselves. Players knew that they would never meet the same person in the reciprocal role. The interactions were anonymous, but the potential donors were shown the history of giving or not giving of the potential recipient before they were asked for their decision. There were six rounds in each group, and each player was once per round a potential donor, and twice per round a receiver, although this was not announced beforehand. The frequency of giving ranged from 48 to 87 percent, depending on the group. As expected, those groups with a lower cost of giving (or with a higher starting account) donated more often. The image score of potential recipients correlated well with their expectation to actually receive money. The amount of discrimination was higher among those players who donated less often:

apparently, those who were more generous cared less about the recipient's image score.

In a similar experiment, Seinen and Schram (2001) found corresponding results. In particular, they concluded that subjects are much more likely to help if they know that their score is passed on. They also found that groups develop different norms, i.e. minimal thresholds for the score. 'Finding a norm that is consistent with the own social status... is important in synchronizing norms within a group'. Seinen and Schram also found clear evidence that the own score becomes an important factor in the decision to help, when players know that it is communicated to future donors.

In an interesting variant of the indirect reciprocation game, Milinski et al. (2002a) showed that if players were given the opportunity, between rounds, to make a public donation to a charity, the amount of their donation correlated positively with the likelihood that they would receive money, in subsequent rounds, from their co-players.

Bolton, Katok and Ockenfels (2001) performed a variety of experiments with high or low costs ($b/c = 5$ or $= 5/3$) and with three different information conditions: (a) no information, (b) first order information (whether the recipient gave help when last in the role of the donor) and (c) second order information (the recipient's decision when last in the role of the donor, and the previous move of the recipient of that game). We note that (b) and (c) both allow SCORING to be implemented, but that (c) does not provide all the information needed to implement a STANDING strategy. The hypothesis that more information leads to more giving is confirmed in the experiments of Bolton et al. (2001). Also, giving is higher in earlier rounds, when reputation has a higher impact on the future income. However, it appears that even if there is no information, some players are prone to cooperate. Furthermore, the decisions of the donors seem also to be affected by how often they were given. This shows that some relevant aspects of the game have not yet been covered by models. Players seem to be affected by what they have received, and tend to give because they received help. Strategies basing the decision to give on the score and, additionally, on the payoff history, i.e. the donor's past income, seem plausible, but apparently have not yet been investigated. But let us stress that Bolton et al. (2001) found a significant positive correlation between the number of gifts given by players and the number they receive. They also found that there is a slight, negative correlation between the number of gifts given and the total payoff obtained by a player.

This last result stands in contradiction to the findings obtained by Wedekind and Braithwaite (2002): in their experiment, those who gave much ended up with the highest payoff. Donors knew only the score of the recipient, calculated according to the SCORING rule, on a scale of integers ranging from -6 to $+6$. Wedekind and Braithwaite found evidence for the OR module. From the twelfth round onward, there was a positive correlation between image score and total payoff, statistically significant in most rounds. Thus generosity pays in this kind of game, which argues for its selective advantage. The correlation within a population increases with the mean generosity of the group. In a subsequent game of direct reciprocity (six rounds of the Prisoner's Dilemma game between the same two players) the display of the previous end score tended to boost cooperation towards generous players in the first three rounds, and then was superseded, reasonably enough, by the personal experience obtained with the given co-player.

A similar interconnection between direct and indirect reciprocation can be found in Milinski et al. (2002b). They combine rounds of an indirect reciprocity game with rounds of a public good game. The donors in the indirect reciprocity game are also informed about the recipients' actions in the public goods game. If the two games alternate, contributions to the public good game remain high, while they quickly deteriorate in an unbroken succession of public good games. This experiment provides evidence that indirect reciprocity has a similar impact as direct reciprocity. Moreover, the version with alternating rounds can be viewed as a sequence of public good games with the possibility, after each round, of rewarding contributors. It thus offers an intriguing complement to the literature on public goods with punishment (see, e. g., Fehr and Gächter 2000).

Engelmann and Fischbacher (2002) ran experiments designed to find out whether donors were more motivated with keeping up their own score or with reacting to the recipients' score. At any time, only half of the players had a public score (assessed according to SCORING), which was displayed when they were recipients. There was clear evidence that donors without score react to the recipient's score; such donors cannot be guided by selfish motives. On the other hand, the propensity to give more than doubled, for many players, if they were told that their action would affect their own score. Such subjects also seem to be less influenced by the recipient's score. This provides strong evidence for selfish reputation-building. Further evidence for such 'strategic' use of reputation has been obtained by Semmann et al. (2004).

In another series of experiments, Milinski et al. (2001) addressed the question of STANDING versus SCORING. Each group included a bogus player who always refused to help. Discriminating players should always refuse to give aid to such a player. The question was: would these players, in turn, be penalised by their co-players or not? The former outcome would speak for the prevalence of a SCORING strategy, the latter for STANDING. Players were again anonymous, and were given, not only the history of the receiver, but also that of the receivers' previous receivers, so that they could judge whether a defection by the receiver was justified or not. It was found that the potential donors of the sham defector (whose refusals were justified) experienced significantly more defections than STANDING would predict, but less than SCORING would predict. Interestingly, the donors of the sham defector tended to be more generous in their other interactions, as if they expected to be punished and wanted to redress their score. This suggests that players do not expect that other players follow a STANDING strategy.

The same result held, surprisingly, when the experimenters provided only the history of the receiver (so that a STANDING strategy was actually impossible to implement). Indeed, the statistics of the games with full information (where donors were provided with the complete histories of all co-players) and with restricted information (where players were only provided with the list of previous actions of the potential recipient) look remarkably similar. With full information, the players took a longer time to reach their decision. This suggests that they tried to interpret the complex histories. But after three or four rounds, it becomes rather complicated to work back through the histories of the recipient's recipients etc., so that players most likely were overburdened, cognitively, and simply stopped to care about details, possibly falling back to some mixture of SCORING and STANDING.

This cognitive problem is, in part, due to the design of the economic experiments. The players do not have a close acquaintance with each other, and can distinguish their group members just by their pseudo-names, so that they are not really involved with them. It could be argued that in more life-like interactions within a real group, individuals are familiar with each other's personalities, and thus find it easier to update their image scores in real time. It would facilitate the players' task of keeping track of their co-players' standing if they were told, after each round, to update all the image scores within the group, and note them down. The drawback of such an instruction is that it necessarily suggests to the participants that these image scores form a key element of the game. The players would no longer be 'naive' with respect to the experiment, but approach it with a certain bias. On the other hand, given that it can by now be granted that *some* type of image score is involved in this kind of game, it could be worth trying to provide players with an instruction like: 'Write down, between each round, who did the right thing, in your eyes, and who did not.' From the resulting protocols, it should be possible to find out the assessment modules and, comparing this with the decisions taken by the players, the action modules.

Another possible way of clarifying the situation would be to subject players to very short histories only. For instance, one could start by explaining the rules of the game, and then let groups of six or ten players actually play ten or twelve rounds, sitting face to face with each other, so that they thoroughly understand what they are about. Then, one could separate the players, place each into some cubicle, and tell them that they would now play the same game, with a new group of co-players with whom they could interact only via computer. In reality, they would all be confronted, in the third round, with a fictitious co-player who had given in the first round, but refused to give in the second round against a recipient who had refused to give in the first round. This should disentangle the SCORING vs STANDING issue. It is considered bad form, in economic games, to mislead players. But in view of the importance of the question, this may be considered a white lie. Morally it may not be quite right, but it can help us to better understand morals and their evolution.

Acknowledgement. H. Brandt and K. Sigmund acknowledge support from the Wissenschaftskolleg "Differential Equations" of the Austrian Science Fund FWF. We thank Josef Hofbauer, Martin Nowak and Drew Fudenberg for helpful conversations.

References

1. Alexander, R.D. (1987), The Biology of Moral Systems, New York: Aldine de Gruyter
2. Axelrod, R and Hamilton, W D (1981), The evolution of cooperation, Science 211: 1390–6
3. Berg, J. Dickhaut, J. and McCabe, K. (1995), Trust, Recipocity and Social History, Games and Economic Behavior 10, 122–142
4. Binmore, K G (1992), Fun and Games: a text on game theory, Health and Co, Lexington, Mass
5. Boerlijst, M C, Nowak, M A and K. Sigmund (1997), The logic of contrition, JTB 185: 281–294
6. Bolton, G., Katok, K. and Ockenfels, A., (2004), Cooperation among strangers with limited information about reputation, Journal of Public Economics, in press
7. Bolton, G., Katok, E. and Ockenfels, A (2004), How effective are online reputation mechanisms? An experimental investigation, Management Science, in press
8. Bolton, G.E. and Ockenfels, A. (2000), ERC: a theory of equity, reciprocity and competition, American Economic Review 90: 166–193
9. Boyd, R and Richerson, P J (1989), The evolution of indirect reciprocity, Social Networks 11: 213–236
10. Boyd, R and Richerson, P J (1992), Punishment allows the evolution of cooperation (or anything else) in sizeable groups, Ethology and Sociobiology 113: 171–195
11. Brandt, H. (2004), URL: http://homepage.univie.ac.at/hannelore.brandt/ simulations Interactive java applets for online experimentation on presented simulations
12. Brandt, H. and Sigmund, K (2004), The logic of reprobation: assessment and action rules for indirect reciprocity, JTB, 231: 475–486
13. Brandt, H. and Sigmund, K. (2005), Indirect reciprocity, image scoring, and moral hazard, Proc. Natl. Acad. Sci. USA 102: 2666–2670
14. Camerer, C E (2003) Behavioral Game Theory, Princeton UP
15. Colman, A. M. (1995) Game Theory and its Applications in the Social and Biological Sciences, Oxford: Butterworth-Heinemann.
16. Dawes, R. M. (1980), Social Dilemmas. Ann. Rev. Psychol. 31: 169.
17. Dellarocas, C (2003), Efficiency and Robustness of binary feedback mechanisms in trading environments with moral hazard (working paper, MIT Sloan School of Management)
18. Ellison, G. (1994), Cooperation in the Prisoner's Dilemma with anonymous random matching, Review of Economic Studies, 61: 567–588
19. Engelmann, D. and U. Fischbacher (2002), Indirect reciprocity and strategic reputation-building in an experimental helping game, working paper, Univ of Zürich

20. Fehr, E. and Gächter, S. (2000), Cooperation and punishment in public goods experiments. Am. Econ. Rev. 90: 980
21. Fehr, E. and Gächter, S. (2002), Altruistic punishment in humans. Nature 415: 137
22. Fehr, E. and Fischbacher, U. (2003), The nature of human altruism, Nature 425: 785–791
23. Ferrière, R. (1998), Help and you shall be helped, Nature 393: 517–519
24. Fishman, M.A. (2003), Indirect reciprocity among imperfect individuals, JTB 225: 285–292
25. Fishman, M.A., Lotem, A. and Stone. L. (2001), Heterogeneity stabilises reciprocal altruism interaction, JTB 209: 87–95
26. Fudenberg D and Maskin, E (1986), The folk theorem in repeated games with discounting or with incomplete information, Econometrica 50: 533–554
27. Gintis, H (2000), Game Theory Evolving, Princeton UP
28. Harbaugh, W T (1998), The prestige motive for making charitable transfers, American Economic Review 88: 277–282
29. Hofbauer, J. and Sigmund, K (1998), Evolutionary Games and Population Dynamics, Cambridge UP
30. Kandori, M (1992), Social norms and community enforcement, The Review of Economic Studies 59: 63–80
31. Keser, Claudia (2002), Trust and Reputation Building in e-Commerce, Working paper, IBM Watson Research Center
32. Leimar, O. and Hammerstein, P. (2001), Evolution of cooperation through indirect reciprocation, Proc R Soc Lond B, 268: 745–753
33. Lotem, A., Fishman, M A and Stone, L (1999), Evolution of cooperation between individuals, Nature 400: 226–227
34. Lotem, A., Fishman, M A and Stone L (2002), Evolution of unconditional altruism through signalling benefits, Proc R. Soc London B, 270: 199–205
35. Milinski, M., Semmann, D., Bakker, T.C.M. and Krambeck, H. J. (2001), Cooperation through indirect reciprocity: image scoring or standing strategy? Proc Roy Soc Lond B 268: 2495–2501
36. Milinski, M., Semmann, D. and Krambeck, H.J. (2002a), Donors in charity gain in both indirect reciprocity and political reputation, Proc Lond Soc B 269: 881–883
37. Milinski, M., Semmann, D. and Krambeck, H.J. (2002b), Reputation helps solve the 'Tragedy of the Commons', Nature 415: 424–426
38. Mohtashemi, M. and L. Mui (2003), Evolution of indirect reciprocity by social information: the role of trust and reputation in evolution of altruism, JTB 223: 523–531
39. Nowak, M.A. and Sigmund, K (1990), The evolution of stochastic strategies in the prisoner's dilemma, Acta Appl. Math. 20: 247–265.
40. Nowak, M.A. and Sigmund, K. (1998a), Evolution of indirect reciprocity by image scoring, Nature 282: 462–466
41. Nowak, M.A. and Sigmund, K. (1998b), The dynamics of indirect reciprocity, JTB 194: 561–574
42. Ohtsuki, H (2004), Reactive strategies in indirect reciprocity, JTB 227: 299–314
43. Ohtsuki H, and Iwasa, Y (2004), How should we define goodness? – Reputation dynamics in indirect reciprocity, JTB 231: 107–120
44. Okuno-Fujiwara, M and Postlewaite, A (1995), Social Norms in Matching Games, Games and Economic Behaviour, 9: 79–109

45. Panchanathan, K. and R. Boyd (2003), A tale of two defectors: the importance of standing for evolution of indirect reciprocity, JTB 224: 115–126
46. Pollock, G B and L A Dugatkin (1992), Reciprocity and the evolution of reputation, JTB 159: 25–37
47. Rosenthal, R W (1979), Sequences of games with varying opponents, Econometrica 47: 1353–1366
48. Seinen, I. and Schram, A. (2001), Social status and group norms: indirect reciprocity in a helping experiment, working paper, CREED, Univ. of Amsterdam
49. Semmann D, Krambeck, H J and Milinski, M (2004), Strategic investment in reputation, J Behav. Ecol. Sociobiol. 56: 248–252
50. Sheratt, T N and Roberts, G (2001), The importance of phenotypic defectors in stabilising reciprocal altruism, Behav. Ecol. 12: 313–317
51. Sigmund, K., Hauert, C., and Nowak, M. A. (2001), Reward and punishment. Proc. Natl. Acad. Sci. 98: 10757–10763
52. Sugden, R. (1986), The Economics of Rights, Cooperation and Welfare, Basil Blackwell, Oxford
53. Takahashi, N. and R. Mashima (2004), The importance of indirect reciprocity: is the standing strategy the answer? Working Paper Hokkaido Univ
54. Trivers, R (1971), The evolution of reciprocal altruism, Quart Rev Biol 46, 35–57
55. Wedekind, C and Milinski, M (2000), Cooperation through image scoring in humans, Science 288, 850–852
56. Wedekind, C (1998), Give and ye shall be recognised, Science 280: 2070–2071
57. Wedekind, C. and Braithwaite, V.A. (2002), The long-term benefits of human generosity in indirect reciprocity, Curr Biol. 12: 1012–1015
58. van der Heijden, E C M (1996), Altruism, Fairness and Public Pensions, PhD thesis Amsterdam
59. Zahavi, A (1995), Altruism as a handicap – the limitations of kin selection and reciprocity, J Avian Biol 26: 1–3

4

The Effects of Migration on Persistence and Extinction

Jingan Cui and Yasuhiro Takeuchi

Summary. The interrelationship between organisms and the environment is essential to the stability or permanence of an ecological system, and the effect of migration on the possibility of species coexistence in an ecological community has been an important subject of research in population biology. Numerous types of models have been proposed and have been used to describe movement or dispersal of population individuals among patches. Some of the existing models deal with a single population dispersing among patches and others deal with predator-prey and competition interactions in patchy environments. Most previous models are based on autonomous ordinary differential equations. Recently, some authors have also studied the influence of migration on time-dependent population models.

In this chapter, we attempt to review key related research and introduce a set of new results for time-dependent population models in patchy environments. We consider a single-species model described by a set of autonomous ordinary differential equations or non-autonomous equations with periodic functions or with dispersal time delays. Also, we consider an age-structure model with or without dispersal delays. Further, we discuss predator-prey or competitive models described by autonomous or time-dependent ordinary differential equations.

4.1 Introduction

The concept of metapopulation is widely used by modelers to explore the effects of spatial heterogeneity on population dynamics. Metapopulation describes a 'population' consisting of many local populations, analogously to a local population consisting of individuals. One kind of basic metapopulation models is the so-called two-population model, which is an extension of traditional single-population models for two local populations connected by migration. These models are deterministic at the population level, and the main issue is the influence of migration on local dynamics. A two-population model can, of course, be extended to many (n) populations connected by migration, though at the cost of not obtaining much meaningful information. Most metapopulation models used by conservationists are n-population dynamic models. A general two-population model takes the form (Hanski 1999):

$$\dot{x}_1 = g_1(x_1)x_1 - \gamma_{12}(x_1)x_1 + \gamma_{21}(x_2)(1 - \delta_1)x_2$$
$$\dot{x}_2 = g_2(x_2)x_2 - \gamma_{21}(x_2)x_2 + \gamma_{12}(x_1)(1 - \delta_2)x_1 \,, \quad (1)$$

where $g_i(x_i)$ is the per capita rate of change of population i due to births and deaths, $\gamma_{ij}(x_i)$ is the per capita rate of emigration from population i to population j, and δ_i is the fraction of migrants dying during migration. The two populations and the respective habitat patches might or might not be similar. In source-sink metapopulations, there are substantial differences between the intrinsic growth rates of local populations.

Since the interrelationship between organisms and the environment seems to be essential to the stability of an ecological system, the effect of migration on the possibility of species coexistence in an ecological community is an important aspect of population biology. Many kinds of models have been proposed and have been used to describe movement or dispersal of population individuals among patches. Some existing models deal with a single population dispersing among patches (Allen 1983, 1987; Amarasekare 1998; Beretta et al. 1987; Cui et al. 2000 ; Gruntfest et al. 1997; Gyllenberg et al. 1997; Hanski, 1999; Hastings, 1983; Holt, 1985; Hui et al. 2005; Levin, 1974; Lu et al. 1993; Mahbuba et al. 1994; Skellam, 1951; Takeuchi, 1989, 1996; Teng et al. 2001; Wang et al. 1997; Zhang et al. 1996). Others deal with predator-prey and competition interactions in patchy environments (Beretta and Solimano 1987; Cui 2002; Cui and Chen 1999, 2001; Cui and Takeuchi 2005; Freedman and Waltman 1977; Hastings 1977; Jansen and Lloyd 2000; Kuang and Takeuchi 1994; Namba et al. 1999; Song and Chen 1998a, 1998b; Takeuchi 1986, 1989; Takeuchi et al. preprint; Teng and Chen 2003; Xu and Chen 2001; Zeng et al. 1994). Most previous models are autonomous ordinary differential equations. Recently, some authors have also studied the influence of migration on time-dependent population models.

In this chapter, we attempt to review key related research and introduce new results for time-dependent population models in patchy environments. In Sect. 4.2, we consider a single-species model described by a set of autonoumous ordinary differential equations or non-autonoumous equations with periodic functions or with dispersal time delays. Also, we consider an age-structure model with or without dispersal delays. In Sect. 4.3, we describe predator-prey models based on autonomous or time-dependent ordinary differential equations. Finally, in Sect. 4.4, competitive models are discussed.

4.2 Single-species system

4.2.1 Population size

How does migration affect the population sizes? This is a basic but important issue. As a standard and not too unrealistic specific example, let us assume

that local dynamics are given by the logistic model, and that emigration is density-independent and the same in the two populations. Equation (1) then becomes:
$$\dot{x}_1 = r_1 x_1 (1 - x_1/k_1) - mx_1 + m(1-\delta)x_2 \\ \dot{x}_2 = r_2 x_2 (1 - x_2/k_2) - mx_2 + m(1-\delta)x_1 \,, \qquad (2)$$
where r_i and k_i are the intrinsic rate of population increase and the carrying capacity of population i respectively, and m is the constant emigration rate.

For positive r_i and k_i, the two-population metapopulation has two equilibria, of which the one corresponding to metapopulation extinction, $(\bar{x}_1, \bar{x}_2) = (0,0)$, is unstable, and the one with $\bar{x}_1 > 0$ and $\bar{x}_2 > 0$ is stable ($\bar{x}$ denotes the equilibrium value). The pooled size of two populations with migration will reduce the overall metapopulation size. The pooled metapopulation size is reduced even if there is no mortality during migration, provided that there is a difference between the two carrying capacities— more individuals move from the population with a larger carrying capacity to that with the smaller one, where their per capita reproductive success is lower than in their parent population. Therefore, random migration of the type assumed in (2) is not expected to evolve by natural selection (Hastings 1983; Holt 1985), implying that migration in natural populations has evolved in response to some forces not considered in this model. The equilibrium population size in the patch with small carrying capacity might, of course, be higher with than without migration, even if there is substantial mortality during migration.

The logistic model assumes that the per capita growth rate decreases linearly with increasing population size. This is a convenient mathematical assumption, but perhaps unrealistic in biology; for instance, density dependence might become strong only when population size is close to the carrying capacity, as would happen when growth is limited by space for individuals' territories. Such situations can be modelled by raising the term x/k in the logistic model to power θ, $(x/k)^\theta$. Assuming that θ is small (< 1), which corresponds to weak density dependence near the carrying capacity, has the interesting consequence that the two populations connected by migration tend to become similar in size, even if the respective carrying capacities are quite different. With increasing mortality during migration, metapopulation size decreases because of the generally reduced growth rate, and with a high rate of mortality during migration the metapopulation might become extinct. The message here is that nonlinearity per capita density dependence makes it ever harder to use observed population sizes or densities to draw inferences about the quality of the respective habitat patches. Assuming that in one population density dependence is weak near the equilibrium ($\theta < 1$) and that the other population has a large carrying capacity, leads to the surprising result that the pooled size of the metapopulation may exceed the sum of the two carrying capacities, even if there is mortality during migration. This happens because one of the populations, the one with the greater carrying capacity, feeds large numbers of migrants to the other population, where weak density dependence allows population density to persist at a high level, much higher than the local carrying capacity. In other words, population sizes in this case are biased towards the k of the population with stronger density dependence, and if this population has larger k, metapopulation size might exceed $k_1 + k_2$ (Holt 1985).

4.2.2 Stability, critical patch number

The n-population model, n populations connected by migration, has been well studied. By applying cooperative system theory (see Smith 1986), Lu and Takeuchi (1993) considered a single-species diffusion system described by

$$\dot{x}_i = x_i f_i(x_i) + \sum_{j \neq i} D_{ij}(x_j - \alpha_{ij} x_i), \quad i = 1, 2, \ldots, n. \tag{3}$$

Here, n is the number of patches, x_i represents the population density of the species in the i-th patch and $f_i(x_i)$ is the per capita rate of change of the population in this patch. D_{ij} is a non-negative diffusion coefficient from patch j to patch i and $D_{ii} = 0$ ($i = 1, 2, \ldots, n$). The parameter $\alpha_{ij} > 0$ corresponds to the boundary conditions of the continuous diffusion case, $\alpha_{ij} = 1$ for Neumann condition, $\alpha_{ij} \neq 1$ for Dirichlet or Robin conditions (Allen 1983). Further, dispersal by linear diffusion implies that the species can move to the interconnected patches with equal probability (Allen 1983).

Define a matrix $D = (d_{ij})$ satisfying

$$d_{ij} = \begin{cases} D_{ij} & \text{for } i \neq j, \\ -\sum_{k=1}^{n} D_{ik} \alpha_{ik} & \text{for } i = j, \end{cases}$$

and assume that D is an irreducible matrix, which implies that the species can reach any i-th patch from any j-th patch. Note that system (3) and D are cooperative (see the definition for cooperative matrix given below).

A matrix $A = (a_{ij})$ is called *cooperative* if all its off-diagonal elements are non-negative, that is, if $a_{ij} \geq 0$ for $i \neq j$. The spectrum of matrix A, written as $\delta(A)$, is the set of eigenvalues of A. Define the stability modulus of A, $s(A)$, as

$$s(A) = \max \{\text{Re}\lambda : \lambda \in \delta(A)\}.$$

First, we consider the critical patch number of system (3). Define the following two matrices

$$A_r = D + \text{diag}(r_1, \ldots, r_n),$$
$$A_f = D + \text{diag}(f_1(0), \ldots, f_n(0)),$$

where $r_i = \sup_{x_i \geq 0} f_i(x_i)$.

4 The Effects of Migration on Persistence and Extinction

Theorem 1. *(Lu and Takeuchi, 1993) Consider system (3).*
(i) *If $s(A_r) < 0$, then $\lim_{t \to \infty} x_i(t) = 0$ for $i = 1, 2, \ldots, n$;*
(ii) *If $s(A_f) > 0$, then $\liminf_{t \geq 0} x_i(t) \geq \delta > 0$ for $i = 1, 2, \ldots, n$, where δ is independent of initial data.*

Let us consider the system described by

$$\dot{x}_i = x_i f(x_i) + D(x_{i+1} - x_i) + D(x_{i-1} - x_i), \quad i = 1, 2, \ldots, n, \quad (4)$$

where D is positive and $x_0 = x_{n+1} = 0$. This system is considered in Allen (1987). Note that (4) has an identical within-patch dynamics. Suppose that $f(x_i) = a$ (exponential growth) or $f(x_i) = a - bx_i$ (logistic growth) ($i = 1, 2, \ldots, n$), where a and b are positive constants. By using Theorem 1, we have

Corollary 1. *Consider system (4) with $f(x_i) = a$ or $f(x_i) = a - bx_i$.*
(i) *If $a < 2D\left(1 + \cos\frac{n\pi}{n+1}\right)$, then $\lim_{t \to \infty} x_i(t) = 0$ for $i = 1, 2, \ldots, n$;*
(ii) *If $a > 2D\left(1 + \cos\frac{n\pi}{n+1}\right)$, then $\liminf_{t \geq 0} x_i(t) \geq \delta > 0$ for $i = 1, 2, \ldots, n$, where δ is independent of initial data.*

By Corollary 1, the critical number of patches for system (4) with an exponential growth or a logistic growth is

$$n_c = \frac{\arccos(a/(2D) - 1)}{\pi - \arccos(a/(2D) - 1)}.$$

This means that fewer patches than n_c imply extinction of the species whereas more patches imply survival.

Now let us consider the global behavior such as extinction of the species and global stability of a positive equilibrium of system (3). It is assumed that $f_i(x_i)$ satisfies a general logistic form, that is, for each $i = 1, 2, \ldots, n$,

$$f_i(0) > 0, \quad \frac{df_i}{dx_i} < 0 \quad \text{for} \quad x_i > 0. \quad (5)$$

Define matrix A as

$$A = D + \text{diag}(f_1(0), \ldots, f_n(0)).$$

Note that $A_r = A_f = A$ by (5).

Theorem 2. *(Lu and Takeuchi 1993) Consider (3) satisfying (5).*
(i) *If $s(A) \leq 0$, then $x = 0$ is globally stable;*
(ii) *If $s(A) > 0$, then one of the following two holds:*
  (a) $\lim_{t \to T_x} x(t) = \infty$ *for every* $x(0) \in R_+^n$, *or*
  (b) $\lim_{t \to \infty} x(t) = x^* > 0$ *for every* $x(0) \in R_+^n$.

Here, $(0, T_x)$ is the maximal interval of existence for $x(t)$ and x^* is a positive equilibrium point of system (3).

This theorem implies that $x = 0$ is globally stable even if $s(A) = 0$. Compare it with Theorem 1. Note also that the theorem classifies the dynamical behavior for system (3), which was shown to be ultimately bounded from below by Theorem 1 (ii). Note that Theorem 2 has assumption (5), which is not supposed in Theorem 1.

Corollary 2. *Consider system (4) with $f(x_i) = a - bx_i (a > 0, b > 0)$,*

(i) *If $a \leq 2D\left(1 + \cos\frac{n\pi}{n+1}\right)$, then $x = 0$ is globally stable;*

(ii) *If $a > 2D\left(1 + \cos\frac{n\pi}{n+1}\right)$, then system (4) possesses a globally stable positive equilibrium point x^*.*

Finally, let us consider the system described by (3) with Neumann boundary condition, that is, with $\alpha_{ij} = 1 (i, j = 1, 2, \ldots, n)$:

$$\dot{x}_i = x_i f_i(x_i) + \sum_{j \neq i} D_{ij}(x_j - x_i), i = 1, 2, \ldots, n. \quad (6)$$

Suppose that

(H1) All solutions of the initial value problem for (6) exist, are unique and are continuable for all positive time;

(H2) $f_i(0) > 0$, $df_i(x_i)/dx_i < 0$ for $x_i > 0$ and $f_i(x_i) < 0$ as $x_i \to +\infty$, $i = 1, 2, \ldots, n$.

Theorem 3. *(Takeuchi 1989) System (6) always has a positive and globally stable equilibrium point, provided assumptions (H1) and (H2) hold.*

When $D_{ij} = 0$ for $i, j = 1, 2, \ldots, n$, (6) gives the dynamics of isolated n patches. By assumption (H2), each isolated patch has a positive and globally stable equilibrium point. Hence, Theorem 3 implies that any diffusion cannot change global stability of system (3). This result greatly generalizes some previous results (cf. Kuang and Takeuchi 1994).

4.2.3 Allee effects

When the population is small, or sparse, reproduction tends to be inhibited. The resulting phenomenon of a positive density dependence in population growth rate at low densities (Allee et al. 1983) is known as the Allee effect. The most obvious, and perhaps the most universal cause of the Allee effect is the increased cost of mate finding at low densities, necessarily experienced by biparentally reproducing organisms.

In order to determine conditions which allow the Allee effect (caused by biparental reproduction) to conserve and create spatial heterogeneity in population densities, Gyllenberg and Hemminki (1997) developed a deterministic model of a symmetric two-patch metapopulation. The subpopulations are assumed to have basically independent dynamics, but their interference can occur both by migration and competition. Let x_i denote the population density in patch i. They arrive at the metapopulation model

$$\dot{x}_i = \beta(x_i)x_i - \mu x_i - \gamma x_i(x_i + sx_j) + D(x_j - x_i), \quad i,j = 1,2, i \neq j. \quad (7)$$

The death rate μ and the migration rate D are the same for both patches. The per capita rate of removal due to competition in patch i is assumed to be proportional (with proportionality constant γ) to $x_i + sx_j (j \neq i)$. The parameter s reflects the degree of inter-patch competition and, normally, $0 \leq s \leq 1$. If $s = 0$, competition occurs only within a patch. If $s > 1$, inter-patch competition is stronger than intra-patch competition, which is biologically quite unrealistic. Competition embraces various types of negative interactions between the subpopulations, including apparent competition. Note that if $D = 0$, the model can be interpreted as a two-species competition model. Accordingly, the between-patch competition is incorporated in the way widely used in mathematical formulation of between-species competition, going back to the classical work of Lotka and Volterra (Murray 1993). In turn, the special case $s = 0$ is analogous to Gruntfest et al. (1997) and Amarasekare (1998). The Allee effect is introduced by assuming β, the per capita birth rate, to represent a function of population density.

Assume first that $\beta(x_i) = \beta$. After rescaling of system (7), the metapopulation model can be written as

$$\dot{x}_i = bx_i - x_i(x_i + sx_j) + D(x_j - x_i), \quad i,j = 1,2, i \neq j, \quad (8)$$

where $b = \beta - \mu$. System (8) can have at most four equilibria:

$$E_0 = (0,0), E_1 = (e_1, e_1), E_2 = (e_2, e_3). E_3 = (e_3, e_2),$$

where

$$e_1 = \frac{b}{s+1}, \quad e_2 = \frac{w - \sqrt{w(w+4D)}}{2(1-s)}, \quad e_3 = \frac{w + \sqrt{w(w+4D)}}{2(1-s)},$$
$$w = (1-s)(b-2D).$$

If the birth rate exceeds the death rate, i.e., if $b > 0$, then E_0 is unstable for all parameter values. This implies that the metapopulation will never go extinct.

If initial population densities are unequal, two kinds of behavior are possible depending on the values of D and s. If $D > b(s-1)/2(s+1)$, then the asymmetric steady state E_2 and E_3 do not exist and the symmetric equilibrium E_1 is stable. Thus, any initial differences between the densities of subpopulations will disappear and we have symmetric coexistence.

Note, however, that the situation is more complicated if the inter-patch competition is stronger than the intra-patch competition, $s > 1$, a case of questionable biological relevance. Now, if the migration rate is sufficiently weak, $D < b(s-1)/2(s+1)$, the asymmetric equilibria E_2 and E_3 exist and are stable, and the symmetric equilibrium E_1 is unstable.

For biparental reproduction, we assume $\beta(x_i) = \beta x_i/(H+x_i)$. To facilitate the analysis, we reduce the number of parameters of the system (7) by scaling. System (7) becomes

$$\dot{x}_i = \frac{\beta x_i^2}{x_i+1} - x_i - \gamma x_i(x_i + s x_j) + D(x_j - x_i), \quad i,j=1,2, i\neq j. \quad (9)$$

At most, three symmetric equilibria exist:

$$E_0 = (0,0), \quad E_1 = (z_1, z_1), \quad E_2 = (z_2, z_2),$$

where

$$z_1 = \frac{\bar{\alpha} - \sqrt{\bar{\alpha}^2 - 4\bar{\gamma}}}{2\bar{\gamma}}, \quad z_2 = \frac{\bar{\alpha} + \sqrt{\bar{\alpha}^2 - 4\bar{\gamma}}}{2\bar{\gamma}},$$

$$\bar{\gamma} = \gamma(s+1), \quad \bar{\alpha} = \beta - \bar{\gamma} - 1.$$

The equilibrium E_0 is asymptotically stable and E_1 is unstable for all positive parameter values. The equilibrium E_2 is stable if, and only if $D > h(z_2) = \frac{1}{2}[\beta z_2(z_2+2)/(z_2+1)^2 - 1 - 2\gamma z_2]$. Gyllenberg and Hemminki (1997) have shown that asymmetric coexistence is possible.

The stability of the equilibrium E_0 implies that the metapopulation will go extinct if the population densities in both patches are sufficiently low. This result differs from the uniparental case.

4.2.4 Periodic system

So far, we have assumed that every population lives in a suitable local habitat, in the sense that the population persists in its local patch in the absence of emigration.

In nature, however, this is not always the case for the actual living environment of endangered species. Because of ecological effects of human activities and industry, e.g., the location of manufacturing industries, pollution of the atmosphere, rivers, and soil, ever more habitats have been broken up into patches and many of these patches are polluted. In some patches, even in all patches, the species will go extinct without a contribution from other patches, and hence the species live in a weak patchy environment. The living environments of some endangered and rare species such as the giant panda (Xun 1994; Yange 1994; Yucun 1994) and *Alligator sinensis* (Zhou 1997) are some convincing examples.

In order to protect endangered and rare species, we have to consider the effects of habitat fragmentation and diffusion on the permanence and extinction of single and multiple species living in weak environments. The present section focuses on the following interesting problem: to what extent does dispersal lead to permanence or extinction of endangered single species which can not persist within isolated patches. The research on this problem firstly appeared in 1998 (see Cui and Chen 1998). Then, Teng and Lu (2001) and Cui and Takeuchi (2004) have recently further investigated this aspect.

Consider

$$\dot{x}_i = x_i[a_i(t) - b_i(t)x_i] + \sum_{j=1}^{n}(1 - \lambda_{ij}(t))D_{ij}(t)x_j - \sum_{j=1}^{n} D_{ji}(t)x_i, \quad (10)$$

$$i = 1, 2, \ldots, n,$$

where x_i $(i = 1, 2, \ldots, n)$ denotes the species number in patch i. $b_i(t), a_i(t)$, $\lambda_{ij}(t)$ and $D_{ij}(t)$ are all continuous functions of time $t \in (-\infty, +\infty)$ and are assumed to be periodic with a common period $\omega > 0$. $a_i(t)$ is the intrinsic growth rate for species in patch i; $b_i(t)$ represents the self-inhibition coefficient and is assumed to be positive for $0 \le t < \omega$; $\lambda_{ij}(t)$ expresses the loss for the species in the process of movement from patch j to patch i and satisfies $0 \le \lambda_{ij}(t) \le 1$, and $D_{ij}(t)$ is the dispersal non-negative coefficient of the species from patch j to patch i.

Let C denote the space of all bounded continuous functions $f: R \to R$, C_+^0 is the set of non-negative $f \in C$ and C_+ is the set of all $f \in C$, such that f is bounded below by a positive constant. Given $f \in C$, we denote

$$f^M = \sup_{t \ge 0} f(t), \quad f^L = \inf_{t \ge 0} f(t)$$

and define the lower average $A_L(f)$ and upper average $A_M(f)$ of f by

$$A_L(f) = \liminf_{r \to \infty, t-s \ge r} (t-s)^{-1} \int_s^t f(\tau) d\tau$$

and

$$A_M(f) = \limsup_{r \to \infty, t-s \ge r} (t-s)^{-1} \int_s^t f(\tau) d\tau$$

respectively. If $f \in C$ is ω-periodic, then the average $A_\omega(f)$ of f must be equal to $A_L(f)$ and $A_M(f)$, that is

$$A_\omega(f) = A_L(f) = A_M(f) = \omega^{-1} \int_0^\omega f(t) dt.$$

In general, the system of differential equations

$$\dot{x} = F(t, x), \quad x \in R^n$$

is said to be permanent if there exists a compact set K in the interior of $R_+^n = \{(x_1, x_2, \ldots, x_n) \in R^n \mid x_i \ge 0, i = 1, 2, \ldots, n\}$, such that all solutions starting in the interior of R_+^n ultimately enter K and remain in it.

Theorem 4. *(Cui and Takeuchi 2005) Assume that there is a non-empty subset I of $N = \{1, 2, \ldots, n\}$ such that*

$$D_{ji}^L > 0 \,(i \in I, j \in I, i \neq j), \quad A_\omega(\phi_I(t)) > 0, \tag{11}$$

where

$$\phi_I(t) = \min_{i \in I} \left\{ a_i(t) - \sum_{j \in N-I} D_{ji}(t) - \sum_{j \in I} \lambda_{ji}(t) D_{ji}(t) \right\}, \tag{12}$$

and, for every $j \in N - I$, there exists at least an integer $i_0 \in I$ such that $D_{ji_0}^L > 0$. Then, (10) is permanent.

Note that system (10) includes the possibility of loss for the species during its dispersion among patches. The system considered in Cui et al. (2005) assumed "safe" dispersion (that is, $\lambda_{ji}(t) = 0$ for any $i, j \in N$).

Let us consider the biological meaning of Theorem 4. Remember that $a_i(t)$ is the intrinsic growth rate for the species in patch i and $D_{ji}(t)$ is the diffusion coefficient for the species from patch i to patch j. Hence, $a_i(t) - \sum_{j \in N-I} D_{ji}(t) - \sum_{j \in I} \lambda_{ji}(t) D_{ji}(t)$ represents the net increasing rate for the species in patch i (that is, intrinsic growth in patch i minus outflow from patch i to patch $j \in N - I$ minus the loss for the species in the process of movement among patches $i \in I$). The assumption $A_\omega(\phi_I(t)) > 0$ implies that the above rate is strictly positive on average. Note that we do not require that $\phi_I(t) > 0$ for all t $(0 \leq t < \omega)$, that is, $a_i(t)$ can be negative at some time durations in $0 \leq t < \omega$. We call such a patch belonging to I "food-rich". On the contrary, the patch $j \in N - I$ is called "food-poor". Note that $\phi_I(t)$ is defined by $a_i(t) - \sum_{j \in N-I} D_{ji}(t) - \sum_{j \in I} \lambda_{ji}(t) D_{ji}(t)$, not by $a_i(t) - \sum_{j \in N} D_{ji}(t) - \sum_{j \in N} \lambda_{ji}(t) D_{ji}(t)$. Hence, patch i is food-rich in the sense that it can provide outflow only for food-poor patch $j \in N - I$, not necessarily to all patchs $j \in N$.

To be permanent for (10), it is sufficient (besides the existence of food-rich patches $i \in I$) that each food-rich patch be connected with all other patches in I ($D_{ji}^L > 0 \,(i \in I, j \in I, i \neq j)$) and each food-poor patch be connected with at least one food-rich patch ($D_{ji_0}^L > 0 \,(i_0 \in I, j \in N - I)$).

According to (11), (12) and Theorem 4, suitable dispersal between "food-rich" and other patches implies permanence. Note that we do not need to take care of the dispersal movement among the "food-poor" patches. This observation may be useful in planning and controlling ecosystems.

Theorem 5. *(Cui and Takeuchi 2004) Assume that there is a non-empty subset I of N such that*

$$D_{ji}^L > 0 \,(i \in N, j \in N, i \neq j), \quad A_\omega(\phi_I(t)) > 0, \tag{13}$$

then (10) has a unique positive ω-periodic solution $(x_1^*(t), x_2^*(t), \ldots, x_n^*(t))$ which is globally asymptotically stable.

The detailed proof of the theorems can be found in Cui and Takeuchi (2004). Note that for the global asymptotic stability, we assume in Theorem 5 that all patches are connected by dispersion ($D_{ij}^L > 0$ required for all $i, j \in N$, not only for $i, j \in I$).

Here, we focus on the difference between systems (6) and (10). Note that $a_i(t)$ is the intrinsic growth rate for the species in patch i in system (10). $a_i(t)$ can become negative at some time durations in $0 \leq t < \omega$ because the natural birth rate may be less than the natural death rate in some seasons. On the other hand, in system (6), $f_i(0)$, the intrinsic growth rate for the species in patch i, is always positive. So, (10) can be used to describe the poor local living environment of some endangered and rare species but (6) can not be used. Equation (6) can be used only to describe the suitable local living environment for some species.

The dispersal term in (10) is different from that in system (6). In fact, in system (6) the individuals can move only from high-density patch to low-density patch, but there is no such unreasonable restrictive condition in system (10).

Example. Consider the following system with five isolated, discrete patches

$$\begin{aligned}
\dot{x}_1 &= x_1(1 + \sin t - x_1), \\
\dot{x}_2 &= x_2(1/2 + \sin t - x_2), \\
\dot{x}_3 &= x_3(\sin t - x_3), \\
\dot{x}_4 &= x_4(-1 + \sin t - x_4), \\
\dot{x}_5 &= x_5(-1 + \cos t - x_5).
\end{aligned} \quad (14)$$

In isolated patches 3, 4 and 5, the local species will go extinct without the contribution from patches 1 and 2.

Now, let us apply Theorem 4 to the patches, some of which are poor living environments, and obtain schemes to ensure permanence of all patches by connecting them suitably.

Questions. Can we choose connective corridors to rescue some endangered local species from extinction? If we can, how many kinds of schemes can be chosen?

Scheme 1. Let $I = \{1\}$. Connecting patch 1 with the other four patches, we have the following model:

$$\begin{aligned}
\dot{x}_1 &= x_1(1 + \sin t - x_1) + 1/4D(t)(x_2 + x_3 + x_4 + x_5) - D(t)x_1, \\
\dot{x}_2 &= x_2(1/2 + \sin t - x_2) + 1/4D(t)x_1 - 1/4D(t)x_2, \\
\dot{x}_3 &= x_3(\sin t - x_3) + 1/4D(t)x_1 - 1/4D(t)x_3, \\
\dot{x}_4 &= x_4(-1 + \sin t - x_4) + 1/4D(t)x_1 - 1/4D(t)x_4, \\
\dot{x}_5 &= x_5(-1 + \cos t - x_4) + 1/4D(t)x_1 - 1/4D(t)x_5.
\end{aligned} \quad (15)$$

If $A_\omega(D(t)) < 1$, then (15) is permanent from Theorem 4.

Scheme 2. Let $I = \{2\}$. Connecting patch 2 with the other four patches, we have the following model:

$$\begin{aligned}
\dot{x}_1 &= x_1(1 + \sin t - x_1) + 1/4D(t)x_2 - 1/4D(t)x_1 , \\
\dot{x}_2 &= x_2(1/2 + \sin t - x_2) + 1/4D(t)(x_1 + x_3 + x_4 + x_5) - D(t)x_2 , \\
\dot{x}_3 &= x_3(\sin t - x_3) + 1/4D(t)x_2 - 1/4D(t)x_3 , \\
\dot{x}_4 &= x_4(-1 + \sin t - x_4) + 1/4D(t)x_2 - 1/4D(t)x_4 , \\
\dot{x}_5 &= x_5(-1 + \cos t - x_4) + 1/4D(t)x_2 - 1/4D(t)x_5 .
\end{aligned} \tag{16}$$

If $A_\omega(D(t)) < 1/2$, then (16) is permanent from Theorem 4.

Scheme 3. Let $I = \{1, 2\}$. Connecting patch 1 with patches 2, 3 and 4, patch 2 with patch 5, we have the following model:

$$\begin{aligned}
\dot{x}_1 &= x_1(1 + \sin t - x_1) + 1/4D(t)(x_2 + x_3 + x_4) - 3/4D(t)x_1 , \\
\dot{x}_2 &= x_2(1/2 + \sin t - x_2) + 1/4D(t)(x_1 + x_5) - 1/2D(t)x_2 , \\
\dot{x}_3 &= x_3(\sin t - x_3) + 1/4D(t)x_1 - 1/4D(t)x_3 , \\
\dot{x}_4 &= x_4(-1 + \sin t - x_4) + 1/4D(t)x_1 - 1/4D(t)x_4 , \\
\dot{x}_5 &= x_5(-1 + \cos t - x_4) + 1/4D(t)x_2 - 1/4D(t)x_5 .
\end{aligned} \tag{17}$$

In this case

$$\phi_I(t) = \min\{1 + \sin t - 1/2D(t), \quad 1/2 + \sin t - 1/4D(t)\}.$$

If $D(t) < 2$ for $0 \leq t < \omega$, then (17) is permanent from Theorem 4.

From the above discussion, we know that some suitable connections between food-rich patches and food-poor ones can make the system permanent.

4.2.5 Dispersal delay

System (10) describes a time-dependent single-species model with a homogeneous spatial patchy environment, and the species instantaneous dispersal among the patches can make the model permanent if there exist "food-rich" patches with suitable connection to all other patches. From a biological point of view, however, it can be argued that the species would require some time to disperse between patches (Y. Takeuchi et al. preprint).

We consider the following single-species system with dispersal time delay in a patchy environment:

$$\dot{x}_i(t) = x_i(t)[a_i(t) - b_i(t)x_i(t)] + \sum_{j=1}^{n}(D_{ij}(t-\tau)x_j(t-\tau) - D_{ji}(t)x_i(t)) ,$$
$$i = 1, 2, \ldots, n ,$$
$$\tag{18}$$

where $x_i (i = 1, 2, \ldots, n)$ denotes the species in patch i. $a_i(t), b_i(t)$ and $D_{ij}(t)$ are all bounded continuous functions of time $t \in R$. τ is a positive constant, which represents the time for the species to disperse between two patches. $a_i(t)$ is the intrinsic growth rate for the species in patch i; $b_i(t)$ represents the self-inhibition coefficient and is assumed to be positive; and $D_{ij}(t)$ is the dispersal coefficient of the species from patch j to patch i, $D_{ij}(t) \geq 0$, $D_{ii}(t) = 0$, and is supposed to be bounded above.

Note that in model (18) with $\tau = 0$, the dispersal term $D_{ij}(t)x_j(t) - D_{ji}(t)x_i(t)$ is different from the commonly used dispersal term $D_{ij}(x_j(t) - x_i(t))$ (see Amarasekare 1998; Berreta and Takeuchi 1987, 1988; Cui et al. 2000; Gruntfest et al. 1997; Gyllenberg and Hemminiki 1997; Hanski 1999; Hastings 1983; Holt 1985; Lu and Takeuci 1993; Mahbuba and Chen 1994; Namba et al. 1999; Takeuchi 1996, etc.). We adopt a new type of dispersal in order to describe another kind of dispersal movement: there is neither cost nor gain during the dispersal process.

Denote by C_B the Banach space of continuous functions $\varphi(t) : [-\tau, 0] \to R^n_+ = \{x \in R^n | x_i \geq 0, i = 1, 2, \ldots, n\}$ with the norm $\| \varphi \| = \sup_{-\tau \leq t \leq 0} | \varphi(t) |$. We denote the phase space of system (18) to be the Banach space C_B. By the fundamental theory of functional differential equations, we know that for any $\varphi \in C_B$ and $\varphi(0) > 0$, system (18) has a unique non-negative solution $(x_1(t, \varphi), x_2(t, \varphi), \ldots, x_n(t, \varphi))$ starting at $t = 0$ with the initial function φ.

Now, we suppose that (18) is ultimately bounded (this is true from Theorem 6 given later). We say that system (18) is *partially persistent* if there exist some sets $I \subset N = \{1, 2, \ldots, n\}$ such that $\liminf_{t \to \infty} V_I(t, \varphi) > 0$ for all solutions, where $V_I(t, \varphi) = \sum_{i \in I} x_i(t, \varphi)$. Further, system (18) is *partially permanent* if there exist some sets $I \subset N$ and a positive constant δ such that $\liminf_{t \to \infty} V_I(t, \varphi) > \delta$ for all solutions. System (18) is called *permanent* if there are positive constants $\delta_i, i = 1, 2, \ldots, n$ such that $\liminf_{t \to \infty} x_i(t, \varphi) > \delta_i$ for all solutions.

We now state the main results with respect to model (18).

Theorem 6. *Y. Takeuchi et al. (2006)* *There exists a positive constant M, which is independent of any solution $(x_1(t, \varphi), x_2(t, \varphi), \ldots, x_n(t, \varphi))$ of (18) with positive initial conditions, such that*

$$\limsup_{t \to \infty} x_i(t, \varphi) \leq M, i = 1, 2, \ldots, n. \qquad (19)$$

Theorem 6 ensures that the population in each patch is ultimately bounded, which is a prerequisite property for a population model.

Theorem 7. *Y. Takeuchi et al. (2006)*

(A) If there is a positive constant m such that $\liminf_{t \to \infty} x_i(t, \varphi) > m$ for some $i \in N$ and for any φ, then there is a positive constant ρ, which is independent of any positive solution of (18), such that $\liminf_{t \to \infty} x_j(t, \varphi) > \rho$ provided $D^L_{ji} > 0$ $(j \neq i)$.

(B) Assume that there are some $i \in N$ such that $\liminf_{t \to \infty} x_i(t, \varphi) > 0$ for any φ, then $\liminf_{t \to \infty} x_j(t, \varphi) > 0$, provided $D^L_{ji} > 0$ $(j \neq i)$.

(C) Assume that there are some $i \in N$ such that $\limsup_{t \to \infty} x_i(t, \varphi) > 0$ for any φ, then $\limsup_{t \to \infty} x_j(t, \varphi) > 0$, provided $D_{ji}^L > 0$ $(j \neq i)$.

Theorem 7 (A) (or (B)) says that if the species in some patch i is survival in the sense of $\liminf_{t \to \infty} x_i(t, \varphi) > m$ (or $\liminf_{t \to \infty} x_i(t, \varphi) > 0$), then the species in all the patches j connected always to the patch i ($D_{ji}^L > 0$ $(j \neq i)$) are also survival in the same sense. This is natural from a biological point of view, since the model describes the *single* species dispersal. (C) ensures that the species in all the patches connected with a survival patch i, in the sense of $\limsup_{t \to \infty} x_i(t, \varphi) > 0$, are also survival in the same sense. We note no restriction on time delay in (A) through (C). This means that the time delay of the dispersal process has a harmless effect on survivability for the system.

The following two theorems give sufficient conditions for partial permanence, permanence and partial persistence of the model.

Theorem 8. Y. Takeuchi et al. (2006)

(A) If there are some sets $I \subset N$ such that

$$A_L(\phi_I(t)) > 0, \phi_I(t) = \min_{i \in I} \left\{ a_i(t) - \sum_{j \in N-I} D_{ji}(t) \right\}, \quad (20)$$

then there exists a positive constant δ such that

$$\liminf_{t \to \infty} V_I(t, \varphi) > \delta$$

for sufficiently small $D_{ji}(t)(i, j \in I, i \neq j)$ in the sense that

$$A_M(\psi_I(t)) < A_L(\phi_I(t)), \quad (21)$$

where $\psi_I(t) = \max_{i \in I} \left\{ \sum_{j \in I} D_{ji}(t) \right\}$.

(B) If further there exists at least one $i \in I$ such that $D_{ki}^L > 0$ for every given $k \in N - I$ and $D_{ji}(t) > 0$ $(i, j \in I, i \neq j)$, then (18) is permanent, that is, there exists $\delta_i > 0$ such that

$$\liminf_{t \to \infty} x_i(t, \varphi) > \delta_i, \quad i = 1, 2, \ldots, n. \quad (22)$$

Conditions (20) and (21) guarantee the partial permanence of the model. First, let us consider the biological meaning of (20). Recall that $a_i(t)$ is the intrinsic growth rate for the species in patch i and $D_{ji}(t)$ represents the dispersion coefficient for the species from patch i to patch j. Hence, $a_i(t) - \sum_{j \in N-I} D_{ji}(t)$ is the net increasing rate for the species in patch i.

The assumption $A_L(\phi_I(t)) > 0$ implies that the net increasing rate is strictly positive in the lower average. We call such patches belonging to class I "food-rich" again. Conversely, patches in $N - I$ are called "food-poor". (20) means that there exist some food-rich patches in the system. Note that (20) is identical with (11) (now $\lambda_{ji}(t) = 0$ for any $i, j \in N$ in (18)).

Now consider (21). $\sum_{j \in I} D_{ji}(t)$ represents the summation of dispersion from some patch i to any other patches $j \in I$, that is, dispersion among food-rich patches. (21) implies that the net increasing rate for the species belonging to food-rich patches exceeds dispersion among them in the lower average, in the sense that $A_L(\phi_I(t) - \psi_I(t)) > 0$. Compared with Theorem 4 that only the existence of food-rich patches is sufficient for the partial permanence of system (18) without the dispersion time delay (i.e., $\tau = 0$), (21) is an additional condition to show partial permanence for (18) with the dispersion time delay. From a biological point of view, however, it may be natural for the species in food-rich patches to have small dispersion among each other because they do not need a high frequency of dispersion to other patches to get more food.

Theorem 8 (B) implies that partial permanence ensures permanence if each food-poor patch is connected to at least one food-rich patch in the sense $D_{ki}^L > 0$ for $k \in N - I, i \in I$, and if each pair in the food-rich patches is connected. The former requirement may be natural for the survival of the species in food-poor patches. On the other hand, the latter requirement needs some explanation. (20) and (21) do not necessarily imply (22) for each patch in food-rich patches, since the two conditions ensure only for *partial* permanence, and there may exist a patch where the species goes extinct even in food-rich patches.

Next, we consider the partial persistence of system (18). Denote

$$b_I(t) = \min_{i \in I} b_i(t), \quad \bar{b}_I(t) = \max_{i \in I} b_i(t), \quad \bar{\phi}_I(t) = \max_{i \in I} \left\{ a_i(t) - \sum_{j \in N - I} D_{ji}(t) \right\}$$

Theorem 9. *Y. Takeuchi et al. (2006) Assume that $\bar{\phi}_I(t)/b_I(t)$ and $\bar{b}_I(t)$ are bounded above. Moreover, assume that*

$$\liminf_{t \to \infty} \frac{\phi_I(t)}{b_I(t)} > 0, \quad \liminf_{t \to \infty} \bar{b}_I(t) > 0.$$

Then $\liminf_{t \to \infty} V_I(t, \varphi) > 0$, *if the following (i) or (ii) holds:*

(i) $\liminf_{t \to \infty} \dfrac{\phi_I(t)}{\bar{b}_I(t)} > \limsup_{t \to \infty} \psi_I(t)$,

(ii) $\liminf_{t \to \infty} \dfrac{\phi_I(t)}{\bar{b}_I(t)} \leq \limsup_{t \to \infty} \psi_I(t)$ *and* $\tau \limsup_{t \to \infty} \psi_I(t) < 1$.

Boundedness for the function $\bar{\phi}_I(t)/b_I(t)$ and $\bar{b}_I(t)$ is reasonable for biological systems. Conditions $\liminf_{t \to \infty} \phi_I(t)/\bar{b}_I(t) > 0$, $\liminf_{t \to \infty} \bar{b}_I(t) > 0$ imply (20). Note that condition (i) in Theorem 9 is essentially the same as (21) if $\bar{b}(t)$ is ignored. Hence, for (18), the possibility of partial persistence is enhanced when the self-inhibition of the species decreases, that is, when the carrying capacity of food-rich patches is increased. This may be reasonable from the biological viewpoint. Theorem 9 (ii) gives the possibility of partial persistence for (18) when the dispersion of the species among food-rich patches is large. Remember that we require small dispersion in Theorem 8 for partial permanence. Theorem 9 (ii) says that, for large dispersion, the time τ necessary for the movement between food-rich patches must be small in the sense $\tau \limsup_{t\to\infty} \psi_I(t) < 1$.

4.2.6 Age-structured model

In this section, we shall study models for populations structured by age. Cui et al. (2000) have considered an age-structured model for *Rana chensinensis*, which occurs mainly in the north and east of China, particularly in Jilin, and is a well-known rare species that has an important medical value. Normally, the adults of *R. chensinensis* live in forests, and they migrate to water fields for reproduction, because water fields or moist habitats are necessary for the growth of young *R. chensinensis* into mature individuals.

Recently, because of ecological effects of human activities and industry, e. g., the location of manufacturing industries, and the pollution of rivers, ever more living habitats of *R. chensinensis* have been disrupted into patches and breeding areas have been damaged in some of these patches. Finally, in these patches, adult *R. chensinensis* will become extinct without contributions from other patches. In fact, many endangered and rare species – e. g., the Chinese sturgeon (Deng et al. 1997), *Alligator sinesis* (Zhou 1997), *Nipponia nippon* (Wang 1997) – face analogous problems because of the destruction and fragmentation of their habitats. In order to protect these species, Cui et al. (2000) put forward the following problem.

Can the local extinction of species in some patches be avoided by building corridors between the patches and controlling the dispersal rates?

To solve this problem, Cui et al. (2000) suppose that the ecosystem is composed of two isolated patches and occupied by a single species of which the individual members have a life history that takes them through two stages, the immature and the mature. Further, the breeding areas are damaged in patch 2. Let $I_i(t)$ and $M_i(t)$ ($i = 1, 2$) denote the density of the immature and mature population in the i-th patch, respectively. To derive our model equations, the following assumptions are made.

A. The birth rate into the immature population in patch 1 is proportional to the existing mature population with proportionality constant a.
B. The death rate of the immature population in patch 1 is proportional both to the existing immature population and to the square of this with proportionality constants c and b, respectively.

C. The death rate of the mature population in the i-th patch is of a logistic nature, i.e., proportional to the square of the population with proportionality constant $\beta_i > 0, i = 1, 2$.

D. The rate of transition from immature individuals to mature individuals is proportional to the existing immature population with proportionality constant α.

If some corridors between the two patches are built, then the mature individuals can move from one patch to another. Then, Cui et al. (2000) assume further:

E. The net exchange of the mature population from patch j to patch i is proportional to the difference of the concentrations $M_j(t) - M_i(t)$ with proportionality constants $D_{ij} \geq 0, i, j = 1, 2, i \neq j$.

Then, the following dispersal model of single-species growth with non-delayed stage structure is obtained by

$$\dot{I}_1(t) = aM_1(t) - bI_1^2(t) - cI_1(t) - \alpha I_1(t),$$
$$\dot{M}_1(t) = \alpha I_1(t) - \beta_1 M_1^2(t) + D_{12}(M_2(t) - M_1(t)), \quad (23)$$
$$\dot{M}_2(t) = -\beta_2 M_2^2(t) + D_{21}(M_1(t) - M_2(t)).$$

If the populations' physical environment fluctuates periodically, then the coefficients in system (23) are all positive and periodic functions with a common period ω. Hence, Cui et al. (2000) obtain the following system corresponding to system (23).

$$\dot{I}_1(t) = a(t)M_1(t) - b(t)I_1^2(t) - c(t)I_1(t) - \alpha(t)I_1(t),$$
$$\dot{M}_1(t) = \alpha(t)I_1(t) - \beta_1(t)M_1^2(t) + D_{12}(t)(M_2(t) - M_1(t)), \quad (24)$$
$$\dot{M}_2(t) = -\beta_2(t)M_2^2(t) + D_{21}(t)(M_1(t) - M_2(t)).$$

The authors (Cui et al. 2000) assume that these functions $a(t), b(t), c(t)$, $\alpha(t), \beta_1(t), \beta_2(t), D_{12}(t)$ and $D_{21}(t)$ in system (24) are all positive and continuous periodic functions with common period ω. Then, we are able to analyze the effect of dispersal on the species survival in system (24), where there are two cases to consider:

Case A. Without dispersal. That is, $D_{12} = D_{21} = 0$ for system (23) and $D_{12}(t) \equiv D_{21}(t) \equiv 0$ for all $t \in [0, \omega]$ in system (24).

Then, systems (23), (24) become the following, respectively.

$$\dot{I}_1(t) = aM_1(t) - bI_1^2(t) - cI_1(t) - \alpha I_1(t),$$
$$\dot{M}_1(t) = \alpha I_1(t) - \beta_1 M_1^2(t), \quad (25)$$
$$\dot{M}_2(t) = -\beta_2 M_2^2(t).$$

$$\dot{I}_1(t) = a(t)M_1(t) - b(t)I_1^2(t) - c(t)I_1(t) - \alpha(t)I_1(t),$$
$$\dot{M}_1(t) = \alpha(t)I_1(t) - \beta_1(t)M_1^2(t), \qquad (26)$$
$$\dot{M}_2(t) = -\beta_2(t)M_2^2(t).$$

Obviously, $M_2(t) \to 0$ as $t \to \infty$ in both systems (25) and (26). $I_1(t), M_1(t)$ for system (25) will globally converge some positive constants, ω-periodic solution, respectively (Theorem 4 in Cui et al. 2000).

Case B. With dispersal.

Two results are obtained for this case (Cui et al. (2000), Theorem 2, Theorem 6), as follows.

Theorem 10. *Cui et al. (2000) System (23) has a unique positive equilibrium and all trajectories in $R_+^3 \setminus \{0\}$ tend to it.*

Theorem 11. *Cui et al. (2000) If $D_{12}^M(c^M + \alpha^M) - a^L \alpha^L < 0$, then system (24) has a unique ω-periodic solution which is globally asymptotically stable.*

By Theorem 10 and Theorem 11, we know the following: any dispersal enables the species in system (23) to avoid future extinction; in system (24), properly controlling the dispersal coefficient $D_{12}(t)$ will also lead to permanence of the species. Therefore, the dispersal of the species between the patches that had been isolated from each other will be effective for this species to ensure permanence. This shows that, by building corridors between the isolated patches, as done in system (23), (24), to allow the adults species to move from one patch to another for reproduction and other behavior, and by controlling the dispersal rates between the patches, we can avoid the local extinction of species in some patches. Hence, corridors between patches and controlling of dispersal rates play an important role in population conservation.

4.2.7 Age-structured model with time delay

Lu and Chen (2002) construct and analyze a fish species age-structured model with diffusion. Fish is a major renewable resource for the human community. Along with developments in science technology, catches of fish have increased. For example, in China traditional fishing used mainly backward manual production tools and was confined to inshore waters, but now the gradual improvement in the technical efficiency of fishing gear and vessels has radically changed the fishing scenario. With the advent of sophisticated fishing instruments, the fishing grounds have been expanded from inshore to offshore waters to meet the ever-growing human demand, and offshore fishing has even become more intensive than that inshore. It appears that some immature fish is being caught, which would affect the resilience of fish resources in the future, negatively impacting on environmental sustanablility. It has become imperative to ensure scientific management of exploitation of biological resources. Thus, an appropriate policy should be put forward to ensure the sustainability of fish prodution. In fact, some relevant measurements have already been taken in China. For instance, the mesh used for fishing needs to meet specified standards in order to avoid immature populations being harvested; fishing activity is restricted to some months of the year to avoid adversely affecting the reproduction and growth of fish. Therefore, it is necessary to attempt modelling the exploitation of inshore-offshore fish with only a mature population harvested.

Clark (1990) dealt with two preliminary models on an inshore-offshore fishery. Pradhan and Chandhuri (1999) investigated an explicit inshore-offshore fishery model, of which the coupling between the inshore and offshore sub-populations of the fish species took place through diffusion from the offshore to the inshore area, and the inshore area was the breeding place. In Lu and Chen (2002), it is assumed that the fish species can breed inshore well as offshore, but the immature individuals can not disperse between the inshore and offshore because they are too weak. The mature fish is harvested both inshore and offshore while it is forbidden to fish the immature. Further, the authors (Lu and Chen 2002) find that the physical environment for the fish population fluctuates periodically due to seasonal effects, long-term changes in climate, and so on. Based on this assumption a mathematical model is put forward and investigated, and a sustainable harvest policy is formulated.

Let $N(t)$ denote a given population at time t, then the population number that survives from t_1 to t_2 is

$$N(t_2) = N(t_1)e^{-\gamma(t_2-t_1)},$$

where γ is the death rate. Let x_{i1} denote the respective immature biomass of the inshore ($i = 1$) or the offshore ($i = 2$) sub-population of the same fish at time $t(> 0)$; x_{i2} denotes the mature biomass of those at time t. To derive our model equations, the following assumptions are made.

(H_1) The birth rate of the immature population in the inshore (offshore) is proportional to the existing mature population with proportionality constant $\alpha_1(\alpha_2) > 0$. The death rate of the immature population in the inshore (offshore) is proportional to the immature population with proportionality constant $\gamma_1(\gamma_2) > 0$.

(H_2) The immature born at time $t - \tau$ that survive to time t exit from the immature population and enter the mature population.

(H_3) The death rate of the mature population in the inshore (offshore) is of logistic mature, i.e., proportional to the square of the population with proportionality constant $\beta_1(\beta_2) > 0$.

(H_4) The net exchange of the mature population from the offshore (inshore) area to the inshore (offshore) area is proportional to the difference of the concentrations $x_{i2}(t) - x_{j2}(t)$ ($x_{j2}(t) - x_{i2}(t)$) with proportionality constants $D_{ji}(D_{ij}) \geq 0$.

(H_5) Selective harvesting of the two sub-populations is considered on the basis of the CPUE (catch-per-unit-effect) hypothesis (Clark, 1990).

Under the above assumptions, Lu et al. (2002) propose a model to describe the inshore-offshore fishing activity as follows:

$$\begin{aligned}
\dot{x}_{i1}(t) &= \alpha_i x_{i2}(t) - \gamma_i x_{i1}(t) - \alpha_i e^{-\gamma_i \tau} x_{i2}(t-\tau)\,, \\
\dot{x}_{i2}(t) &= \alpha_i e^{-\gamma_i \tau} x_{i2}(t-\tau) - \beta_i x_{i2}^2(t) + D_{ij}(x_{j2}(t) - x_{i2}(t)) - E_i q_i x_i(t)\,, \\
x_{ij}(t) &= \phi_{ij}(t) \geq 0\,, \quad t \in [-\tau, 0]\,, \quad \phi_{ij}(0) > 0\,, \quad i,j=1,2\,,
\end{aligned} \quad (27)$$

where $\alpha_i, \beta_i, \gamma_i, D_{ij}, \tau$ have the definitions as above in (H_1) − (H_4). E_1 is the harvesting effect for the inshore mature population, E_2 for the offshore. q_1 is the catchibility coefficient of the inshore population, q_2 of the offshore. $\phi_{12}(t)$ ($\phi_{22}(t)$) is the given initial mature population in the inshore (offshore), and $\phi_{11}(t)$ ($\phi_{21}(t)$) is the derived initial immature population in the inshore (offshore). For continuity of initial conditions, one requires

$$x_{i1}(0) = \int_{-\tau}^{0} \alpha_i x_{i2}(s) e^{\gamma_i s}\, ds\,. \quad (28)$$

If the population's physical environment fluctuates periodically, then the coefficients in system (27) are all positive and periodic functions with a common period ω. Then, Lu and Chen (2002) obtain the following system (29).

$$\begin{aligned}
\dot{x}_{i1}(t) &= \alpha_i(t) x_{i2}(t) - \gamma_i(t) x_{i1}(t) - \alpha_i(t-\tau) e^{-\int_{t-\tau}^{t} \gamma_i(s)\, ds} x_{i2}(t-\tau)\,, \\
\dot{x}_{i2}(t) &= \alpha_i(t-\tau) e^{-\int_{t-\tau}^{t} \gamma_i(s)\, ds} x_{i2}(t-\tau) - \beta_i(t) x_{i2}^2(t) \\
&\quad + D_{ij}(t)(x_{j2}(t) - x_{i2}(t)) - E_i(t) q_i(t) x_{i2}(t)\,, \\
x_{ij}(t) &= \phi_{ij}(t) \geq 0\,, \quad t \in [-\tau, 0]\,, \quad \phi_{ij}(0) > 0\,, \quad i,j=1,2\,.
\end{aligned} \quad (29)$$

Obviously, $\exp(-\int_{t-\tau}^{t} \gamma_i(s) ds)$ is also a periodic function with period ω. $\phi_{12}(t)(\phi_{22}(t))$ is the initial mature population in the inshore (offshore) area, and $\phi_{11}(t)(\phi_{21}(t))$ is the derived initial immature population in those areas. For continuity of initial conditions, it is required that

$$x_{i1}(0) = \int_{-\tau}^{0} \alpha_i(s) x_{i2}(s) e^{\int_{0}^{s} \gamma_i(\theta) d\theta}\, ds\,. \quad (30)$$

Lu and Chen (2002) obtain:

Theorem 12. *If $\alpha_i(t), \gamma_i(t), \beta_i(t), E_i(t), q_i(t)$ are continuous positive periodic functions with period ω, $E_i^M q_i^M < \alpha_i^L e_i^{-r_i^M}$, then system (29) has a unique positive ω-periodic solution which attracts all positive solutions.*

The results in Lu and Chen (in press) illustrate that the fish production would be sustainable if the harvesting were controlled suitably and only for the mature fish both in inshore and offshore areas. This is a benefit for the policy of sustainability and development. Hence, by prohibiting the harvesting of immature fish species, and through the dispersal of mature species between inshore and offshore areas and control of the harvesting effort, the continuously fluctuating fish environment would be maintained, in the sense that all species would be permanent both in inshore and offshore areas.

4.3 Predator-prey system

Let us now consider predator-prey systems with diffusion. These systems have been considered by many authors (for example, Cui and Chen 2001; Cui and Takeuchi to appear; Freedman and Waltman 1977; Holt 1985; Kuang and Takeuchi 1994; Namba et al. 1999; Song and Chen 1998; Takeuchi 1996; Xu and Chen 2001).

4.3.1 Autonomous predator-prey system

Consider a model with barriers only for prey dispersion (Freedman et al. 1977).

$$\dot{x}_i = x_i g_i(x_i) - y p_i(x_i) - \varepsilon_i h_i(x_i) + \sum_{j=1, j\neq i}^{n} \pi_{ji} \varepsilon_j h_j(x_j) ,$$

$$\dot{y} = y[-s(y) + \sum_{i=1}^{n} c_i p_i(x_i)] \quad i = 1, \ldots, n .$$
(31)

Here, $x_i(t)$ represents the prey population in the i-th patch for $i = 1, \ldots, n$. The barriers are considered only as far as the dispersion of the prey population is considered. The predator population has no barriers between the patches, and $y(t)$ is the total predator population for all n patches.

$g_i(x_i)$ is the specific growth rate for the prey population in the i-th patch. The $g_i(x_i)$ is assumed to be a decreasing function of x_i, eventually becoming negative, since the i-th patch can support only a finite population due to limited resources. That is,

(P1) $g_i(0) > 0, dg_i(x_i)/dx_i < 0$; there is a $K_i > 0$ such that $g_i(K_i) = 0, i = 1, \ldots, n$.

$p_i(x_i)$ is the predator functional response of the predator population to the prey in the i-th patch. Since the predator functional response is an increasing function of prey numbers, we assume that

(P2) $p_i(0) = 0, dp_i(x_i)/dx_i > 0, i = 1, \ldots, n$.

$h_i(x_i)$ represents the pressure or need for the prey population to leave the i-th patch and seek another patch in the environment. Clearly, the pressure to disperse increases with increasing population size. Hence, we assume that

(P3) $h_i(0) = 0, \eta_i \geq dh_i(x_i)/dx_i \geq dh_i(0)/dx_i > 0, i = 1, \ldots, n$.
$\varepsilon_i (i = 1, \ldots, n)$ is an inverse barrier strength. If $\varepsilon_i = 0$, then the prey population may not leave the i-th patch.
π_{ji} is the probability that a given member of the prey population, having left the j-th patch, will arrive safely at the i-th patch. Clearly,

$$0 \leq \pi_{ji} \leq 1, \quad \sum_{j=1, j \neq i}^{n} \pi_{ji} \leq 1, (i, j = 1, \ldots, n; i \neq j).$$

$s(y)$ is the density-dependent death rate of the predator in the absence of prey. Since this is likely to be an increasing function of y (when food is scarce, large populations will compete more rigorously for the food), we assume that

(P4) $s(0) > 0, ds/dy \geq 0$.
c_i is the conversion ratio of prey into predator.
Finally, we assume the following:

(P5) All functions are sufficiently smooth, so that solutions to initial value problems of system (31) exist, are unique and are continuable for all $t > 0$.
Let

$$\varepsilon_i = \alpha_i \varepsilon (\alpha_i > 0), \quad \pi_{ii} = -1, \quad \pi_{ji} \neq 0, \quad i, j = 1, \ldots, n.$$

System (31) becomes

$$\begin{aligned} \dot{x}_i &= x_i g_i(x_i) - y p_i(x_i) + \varepsilon \sum_{j=1}^{n} \pi_{ji} \alpha_j h_j(x_j), \\ \dot{y} &= y[-s(y) + \sum_{i=1}^{n} c_i p_i(x_i)], \quad i = 1, \ldots, n. \end{aligned} \quad (32)$$

Let us first consider equilibrium points of (32). The origin, $E_0 = (0, \ldots, 0)$, is clearly an equilibrium. There may be an equilibrium in the positive x subspace, that is, of the form $\bar{E} = (\bar{x}_1, \ldots, \bar{x}_n; 0)$, where $\bar{x}_i > 0$ ($i = 1, \ldots, n$). Note that $\bar{E}$ is a function of ε and $\bar{E} = (K_1, \ldots, K_n; 0)$ at $\varepsilon = 0$. $\bar{E}$ and E_0 are the only possible equilibria with $y = 0$. There may exist a positive equilibrium point.

Theorem 13. *(Freedman and Waltman 1977) The $\bar{E}$ exists for ε satisfying*

$$0 \leq \varepsilon \leq \min(g_i(0)/\alpha_i), \quad i = 1, \ldots, n. \quad (33)$$

Theorem 14. *(Freedman and Waltman 1977) Suppose that $\bar{E}$ does not exist. Then*

$$\lim_{t \to \infty} x_i(t) = 0, \quad i = 1, \ldots, n, \quad \lim_{t \to \infty} y(t) = 0$$

Theorem 15. *(Freedman and Waltman 1977) Let $\bar{E}$ exist uniquely. Further, let $E = (\bar{x}_1, \ldots, \bar{x}_n)$ be a globally stable equilibrium point of the following subsystem of (32):*

$$\dot{x}_i = x_i g_i(x_i) - \varepsilon \alpha_i h_i(x_i) + \varepsilon \sum_{j=1, j \neq i}^{n} \pi_{ji} \alpha_j h_j(x_j). \tag{34}$$

Define

$$d(\varepsilon) = -s(0) + \sum_{i=1}^{n} c_i p_i(\bar{x}_i(\varepsilon)).$$

If $d(\varepsilon) < 0$, then $\lim_{t \to \infty} x_i(t) = \bar{x}_i$ for $i = 1, \ldots, n$, and $\lim_{t \to \infty} y(t) = 0$.

Theorem 16. *(Freedman and Waltman 1977) Let E exist and be globally stable in intR_+^n with respect to (34). Then, system (32) is persistent, provided $d(\varepsilon) > 0$.*

4.3.2 Time-dependent predator-prey system

Consider the following periodic system

$$\begin{aligned}\dot{x}_1 &= x_1[a_1(t) - b_1(t)x_1 - c_1(t)y] + D_{12}(t)x_2 - D_{21}(t)x_1 \\ \dot{x}_2 &= x_2[a_2(t) - b_2(t)x_2] + D_{21}(t)x_1 - D_{12}(t)x_2 \\ \dot{y} &= y[-d(t) + e(t)x_1 - f(t)y]. \end{aligned} \tag{35}$$

Here, x_1 and x_2 are the density of the prey in patch 1 and in patch 2, respectively, and y represents the density of the predator in patch 1. All coefficients in (35) are ω-periodic and continuous for $t \geq 0$, $b_1(t), b_2(t), f(t), D_{12}(t)$ and $D_{21}(t)$ are all positive, and $d(t), a_1(t), c_1(t), e(t)$ are non-negative.
    For (35) we make the following assumptions
    (P6) $A_\omega[a_1(t) - D_{21}(t)] > 0$.

Theorem 17. *(Cui 2002) Under the assumption (P6), system (35) is permanent if, and only if*

$$A_\omega[-d(t) + e(t)x_1^*(t)] > 0.$$

Here, $(x_1^*(t), x_2^*(t))$ is the positive periodic solution of the system

$$\begin{aligned}\dot{x}_1 &= x_1[a_1(t) - b_1(t)x_1] + D_{12}(t)x_2 - D_{21}(t)x_1 \\ \dot{x}_2 &= x_2[a_2(t) - b_2(t)x_2] + D_{21}(t)x_1 - D_{12}(t)x_2. \end{aligned}$$

Note that $a_2(t)$ may be negative on some intervals of $[0, \omega)$.
    Recently, Cui and Takeuchi (2005) obtained a more general result than Theorem 17.

We consider the following predator-prey system in a patchy environment:

$$\dot{x}_1 = x_1[a_1(t) - b_1(t)x_1 - y\varphi(t,x_1)] + \sum_{j=1}^{n}(D_{1j}(t)x_j - D_{j1}(t)x_1)$$

$$\dot{x}_i = x_i[a_i(t) - b_i(t)x_i] + \sum_{j=1}^{n}(D_{ij}(t)x_j - D_{ji}(t)x_i), \quad i = 2,\ldots,n \quad (36)$$

$$\dot{y} = y[-d(t) + e(t)x_1\varphi(t,x_1) - f(t)y].$$

All coefficients in (36) are ω-periodic and continuous for $t \geq 0$. Suppose that

$$\varphi(t,x_1) \geq 0, \frac{\partial}{\partial x_1}\varphi(t,x_1) \leq 0, \frac{\partial}{\partial x_1}(x_1\varphi(t,x_1)) \geq 0, \quad (37)$$

$$D_{ji}^L > 0 (i,j \in N, i \neq j), \quad A_\omega(\phi_I(t)) > 0, \quad (38)$$

where

$$\phi_I(t) = \min_{i \in I}\left\{a_i(t) - \sum_{j \in N-I} D_{ji}(t)\right\}.$$

Theorem 18. *(Cui and Takeuchi 2005) Suppose that there exists a nonempty set $I \subset N = \{1,2,\ldots,n\}$ such that the assumption (38) holds. If*

$$A_\omega\left(-d(t) + e(t)x_1^*(t)\varphi(t,x_1^*(t))\right) > 0, \quad (39)$$

then system (36) is permanent, where $x^*(t) = (x_1^*(t), x_2^*(t), \ldots, x_n^*(t))$ is the globally asymptotically stable positive ω-periodic solution of the system

$$\dot{x}_i = x_i[a_i(t) - b_i(t)x_i] + \sum_{j=1}^{n}(D_{ij}(t)x_j - D_{ji}(t)x_i), \quad (i = 1,2,\ldots,n). \quad (40)$$

If

$$A_\omega\left(-d(t) + e(t)x_1^*(t)\varphi(t,x_1^*(t))\right) \leq 0, \quad (41)$$

then $\lim_{t\to\infty} y(t) = 0$.

Under (38), system (36) without predator $y(t)$ has a unique positive periodic solution which is globally asymptotically stable (see Theorem 5). Theorem 18 says that system (36) with $y(t)$ is permanent under (39) if the prey dispersal system has such a globally asymptotically stable positive ω-periodic solution. Otherwise, the predator goes extinct by Theorem 18.

In (39), the term $e(t)x_1^*(t)\varphi(t,x_1^*(t))$ describes the growth of the predator by foraging the prey in patch 1, of which the quantity is specified as $x_1^*(t)$. Note that $(x_1^*(t), x_2^*(t), \ldots, x_n^*(t))$ is a globally asymptotically stable periodic solution in the prey dispersal system and the predator is confined only to patch 1. Hence, condition (39) implies that growth by foraging minus death for the predator is positive on average. If it is non-positive, then extinction is inevitable for the predator. Note that we assumed (37), of which the last condition implies that the system can be permanent if $x_1^*(t)$ is large. This is reasonable, since the predator is confined to patch 1.

4.4 Competitive system

4.4.1 Autonomous competitive system

Let us consider a two-species competitive Lotka-Volterra system connected by diffusion (Takeuchi 1996), that is, competition between two species (denoted by x and y) over two patches (denoted by 1 and 2) described as follows:

$$\begin{aligned} \dot{x}_i &= x_i(K_i - p_i x_i - q_i y_i) + \varepsilon_i(x_j - x_i) ,\\ \dot{y}_i &= y_i(L_i - r_i x_i - s_i y_i) + \delta_i(y_j - y_i) , \quad i,j = 1,2; i \neq j , \end{aligned} \qquad (42)$$

where x_i and $y_i (i = 1,2)$ are the numbers of species x and y in patch i, and the parameters $K_i, L_i, p_i, q_i, r_i, s_i, \varepsilon_i, \delta_i (i = 1,2)$ are all positive constants.

Clearly, the origin $E_0 = (0,0,0,0)$ is always an equilibrium point. For any $\varepsilon_i > 0$, there exists an equilibrium $E_x = (\bar{x}_1, \bar{x}_2, 0, 0)$ where $\bar{x}_i > 0, i = 1,2$. Similarly, there exists an equilibrium $E_y = (0, 0, \bar{y}_1, \bar{y}_2)$ for any $\delta_i > 0$, where $\bar{y}_i > 0, i = 1,2$. There may exist a positive equilibrium point which is of interest to us when we consider global stability of the system.

The Jacobian matrix of system (42) evaluated at equilibrium E_x is

$$J(E_x) = \begin{pmatrix} J_x(E_x) & -M_x \\ 0 & J_y(E_x) \end{pmatrix},$$

where

$$J_x(E_x) = \begin{pmatrix} K_1 - 2p_1\bar{x}_1 - \varepsilon_1 & \varepsilon_1 \\ \varepsilon_2 & K_2 - 2p_2\bar{x}_2 - \varepsilon_2 \end{pmatrix},$$

and

$$J_y(E_x) = \begin{pmatrix} L_1 - r_1\bar{x}_1 - \delta_1 & \delta_1 \\ \delta_2 & L_2 - r_2\bar{x}_2 - \delta_2 \end{pmatrix}.$$

Similarly, the Jacobian matrix at E_y is

$$J(E_y) = \begin{pmatrix} J_x(E_y) & 0 \\ -M_y & J_y(E_y) \end{pmatrix}.$$

Remember that the stability modulus of A is given by $s(A)$, as

$$s(A) = \max\{Re\lambda : \lambda \in \delta(A)\} .$$

We have the following result.

Theorem 19. *(Takeuchi 1996) If both $s_x = s(J_y(E_x))$ and $s_y = s(J_x(E_y))$ are positive, then system (4.1) is permanent. Furthermore, if a positive equilibrium point is unique, it is globally stable.*

It is easy to extend this theorem to the systems which have three or more patches (see Takeuchi 1996).

4.4.2 Periodic competitive system

Now let us consider the system

$$\dot{x}_i = x_i[a_i(t) - b_i(t)x_i - c_i(t)y_i] + \sum_{j=1}^{n} D_{ij}(t)(x_j - x_i)$$

$$\dot{y}_i = y_i[d_i(t) - e_i(t)x_i - q_i(t)y_i] + \sum_{j=1}^{n} \lambda_{ij}(t)(y_j - y_i) \qquad (43)$$

$$i = 1, 2, \ldots, n.$$

where y_i is the number of species y in patch i, $c_i(t), e_i(t)$, and $\lambda_{ij}(t)$ are all non-negative and bounded continuous functions. In addition, $d_i \in C, q_i \in C_+$ and $\lambda_{ii}(t) \equiv 0$. We will consider the effect of the dispersal and competitive species y on the survival of the native species x.

Theorem 20. *(Cui and Chen 2001) If $(x_1(t), \ldots, x_n(t), y_1(t), \ldots, y_n(t))$ is the solution of (43) with a positive initial condition, then there exist positive constants N_{xi}, N_{yi} and τ_1, such that*

$$0 < x_i(t) \leq N_{xi}, 0 < y_i(t) \leq N_{yi}, \quad i = 1, 2, \ldots, n, \quad t \geq \tau_1.$$

Theorem 21. *(Cui and Chen 2001)*

(I). Suppose that the assumption (C1) or (C2) holds,

(C1) There exists $i_0(1 \leq i_0 \leq n)$, such that $A_L(\mu_1) > 0$, where $\mu_1(t) = a_{i_0}(t) - c_{i_0}(t)N_{yi_0} - \sum_{j=1}^{n} D_{i_0 j}(t)$,

(C2) $A_L(\phi_1) > 0$, where $\phi_1(t) = \min_{1 \leq i \leq n}\{a_i(t) - c_i(t)N_{yi} - \sum_{j=1}^{n} D_{ij}(t) + \sum_{j=1}^{n} D_{ji}(t)\}$.

Then, there exist ζ_{xi} $(0 < \zeta_{xi} \leq N_{xi})$ and $\tau_2 \geq \tau_1$, such that

$$x_i(t) \geq \zeta_{xi} \quad \text{for} \quad i = 1, 2, \ldots, n, \quad t \geq \tau_2$$

(II). Suppose that $\lambda_{ij}(t)(i \neq j)$ are continuous and bounded above and below by positive constants, and the assumption (C3) or (C4) holds.

(C3) There exists $i_0(1 \leq i_0 \leq n)$, such that $A_L(\mu_2) > 0$, where $\mu_2(t) = d_{i_0}(t) - e_{i_0}(t)N_{xi_0} - \sum_{j=1}^{n} \lambda_{i_0 j}(t)$,

(C4) $A_L(\phi_2) > 0$, where $\phi_2(t) = \min_{1 \leq i \leq n} \{d_i(t) - e_i(t)N_{xi} - \sum_{j=1}^{n} \lambda_{ij}(t) + \sum_{j=1}^{n} \lambda_{ji}(t)\}$.

Then, there exist $\zeta_{yi}(0 < \zeta_{yi} \leq N_{yi})$ and $\tau_3 \geq \tau_2$, such that
$$y_i(t) \geq \zeta_{yi} \quad \text{for} \quad i = 1, 2, \ldots, n, \quad t \geq \tau_3.$$

Theorem 22. *(Cui and Chen 2001) Under the assumption (C3) or (C4), if $\int_0^{+\infty} E(t)dt = -\infty$, where*
$$E(t) = \max_{1 \leq i \leq n} \{a_i(t) - c_i(t)\zeta_{yi} - \sum_{j=1}^{n} D_{ij}(t) + \sum_{j=1}^{n} D_{ji}(t)\},$$

then the solution of (43) satisfies
$$\lim_{t \to +\infty} x_i(t) = 0, \quad i = 1, 2, \ldots, n$$

In this section, we have discussed permanence of dispersal models with two competitors. Theorem 19 shows that permanence of (42) is realized if the two boundary equilibria E_x and E_y are unstable. By choosing proper values of dispersal coefficients, the two equilibria can be made unstable. The choices depend on the patch dynamics without diffusion (see Takeuchi 1996 in detail). By contrast, we can not obtain such exquisite results for time-dependent models (43). In Theorem 21, the permanence of (43) is considered. We have obtained some sufficient conditions which depend on the upper bounds of the positive solution of system (43). These results on permanence of (43) have room for improvement.

The results presented in this chapter suggest that dispersal movements of populations among discrete patches is an important factor influencing the dynamics of ecosystems. Dispersal can make the species either permanent or extinct, based on changing dispersal rates.

References

1. Aiello, W.G. and H.I. Freedman (1990), A time-delay model of single-species growth with stage structure, Math. Biosci., **101**: 139–153.
2. Aiello, W.G., H.I. Freedman and J. Wu (1992), Analysis of a model representing stage-structure population growth with state-dependent time delay, SIAM J. Appl. Math. **52**: 855–869.
3. Allee, W. C., A. Emerson, O. Park, T. Park and K. Schmidt (1949), *Principles of Animal Ecology*, (Saunders, Philadelphia).

4. Allen, L.J.S. (1983), Persistence and extinction in single-species reaction-diffusion models, Bull. Math. Biol. **45**: 209–227.
5. Allen, L.J.S. (1987), Persistence, extinction, and critical patch number for island populations, J. Math. Biol. **24**: 617–625.
6. Amarasekare, P. (1998), Interactions between local dynamics and dispersal: Insight from single species species models, Theor. Popul. Biol. **53**: 44–59.
7. Beretta, E. and Y. Takeuchi (1987), Global stability of single-species diffusion Volterra models with continuous time delays, Bull. Math. Biol. **49**: 431–448.
8. Beretta, E. and Y. Takeuchi (1988), Global asymptotic stability of Lotka-Volterra diffusion models with continuous time delays, SIAM J. Appl. Math., **48**: 627–651.
9. Beretta, E. and F. Solimano (1987), Global stability and periodic orbits for two patch predator-prey diffusion delay models, Math. Biosci. **85**: 153–183.
10. Clark, C. W. (1990), *Mathematical Bioeconomics: The Optimal Management of Renewable Resources, 2nd edn.*, (Wiley, New York).
11. Cui, J. (2002), The effect of dispersal on permanence in a predator-prey population growth model, Computers Math. Applic. **44**: 1085–1097.
12. Cui, J. and L. Chen (1998) The effect of diffusion on the time varying Logistic population growth, Computers Math. Applic. **36**: 1–9.
13. Cui, J. and L. Chen (1999), The effect of habitat fragmentation and ecological invasion on population sizes, Computers Math. Applic. **38**: 1–11.
14. Cui, J. and L. Chen (2001), Permanence and extinction in logistic and Lotka-Volterra systems with diffusion, J. Math. Anal. Appl. **258**: 512–535.
15. Cui, J., L. Chen and W. Wang (2000), The effect of dispersal on population growth with stage-structure, Computers Math. Applic. **39**: 91–102.
16. Cui, J., Y. Takeuchi and Z. Lin (2004), Permanence and extinction for dispersal population system, *J. Math. Anal. Appl.* **298**: 73–93.
17. Cui, J. and Y. Takeuchi (2005), Permanence of a single-species dispersal system and predator survival, *J. Comp. Appl. Math.* **175**: 375–394.
18. Deng, X. and Z. Deng (1997), Progress in the conservation biology of Chinese sturgeon, Zoological Research **18(1)**: 113–120.
19. Freedman, H.I. (1987), Single species migration in two habitats: persistence and extinction, Math. Model. **8**: 778–780.
20. Freedman, H.I. and Y.Takeuchi (1989), Global stability and predator dynamics in a model of prey dispersal in a patchy environment, Nonlinear Anal. TMA **13**: 993–1002.
21. Freedman, H. I. and P. Waltman (1977), Mathematical models of population interaction with dispersal. I. Stability of two habitats with and without a predator, SIAM J. Math. **32**: 631–648.
22. Gruntfest, Y., R. Arditi and Y. Dombrovsky (1997), A fragmented population in a varying environment, J. Theor. Biol. **185**: 539–547.
23. Gyllenberg, M. and J. Hemminiki (1997), Habitat deterioration, habitat destruction, and metapopulation persistence in a heterogenous landscape, Theor. Pop. Biol. **52**: 198–215.
24. Gyllenberg, M. and I. Hanski (1999), Allee effects can both conserve and create spatial heterogeneity in population densities, Theor. Pop. Biol. **56**: 231–242.
25. Hanski, I. (1999), *Metapopulation Ecology*, (Oxford University Press).
26. Hastings, A. (1977), Spatial heterogeneity and the stability of predator prey systems, Theor. Pop. Biol. **12**: 37–48.

27. Hastings, A. (1983), Can spatial variation alone lead to selection for dispersal? *Theor. Pop. Biol.*, 24: 244–251.
28. Holt, R. D. (1985), Population dynamics in two-patch environments: some anomalous consequences of optimal habitat distribution, *Theor. Pop. Biol.* **28**: 181–208.
29. Hui, J., and L. Chen (2005), A single species model with impulsive diffusion, Acta Mathematicae Applicatae Sinica. English Series **21**: 43–48.
30. Jansen, V. A. A. and A. L. Lloyd (2000), Local stability analysis of spatially homogeneous solutions of multi-patch systems, J. Math. Biol. **41**: 232–252.
31. Kuang, Y. and Y. Takeuchi (1994), Predator-prey dynamics in models of prey dispersal in two-patch environments, Math. Biosc. **120**: 77–98.
32. Levin, S. A. (1974), Dispersion and population interactions, Amer. Natur. **108**: 207–228.
33. Liu, S., L. Chen and R. Agarwal (2002), Recent progress on stage-structured population dynamics, Mathematical and Computer Modelling **36**: 1319–1360.
34. Lu, Z. and L. Chen (2002), Global attractivity of nonautonomous inshore-offshore fishing model with stage-structure, Appli. Anal. **81**: 589–605.
35. Lu, Z. and Y. Takeuchi (1993), Global asymptotic behavior in single-species discrete diffusion systems, J. Math. Biol. **32**: 67–77.
36. Mahbuba, R. and L. Chen (1994), On the nonautonomous Lotka-Volterra competition system with diffusion, Differential Equations and Dynamical systems **2**: 243–253.
37. Murray, J. D. (1993), *Mathematical Biology*, (Springer, Berlin).
38. Namba, T., A. Umemoto and E. Minami (1999), The effects of habitat fragmentation on persistence of source-sink metapopulation in systems with predators and prey or apparent competitors, Theor. Pop. Biol. **56**: 123–137.
39. Pradhan, T. and K.S. Chandhari (1999), Bioeconomic modelling of selective harvesting in an inshore-offshore fishery. Differential Equations and Dynamical Systems **7**: 305–320.
40. Skellam, J. D. (1951), Random dispersal in theoretical population, Biometrika, **38**: 196–216.
41. Smith, H. L. (1986), Cooperative systems of differential equation with concave nonlinearities, Nonlinear Analysis **10**: 1037–1052.
42. Song, X. and L. Chen (1998a), Persistence and global stability of nonautonomous predator-prey system with diffusion and time delay, Computers Math. Applic. **35**: 33–40.
43. Song, X. and L. Chen (1998b), Persistence and periodic orbits for two species predator-prey system with diffusion, Canad. Appl Math. Quart. **6**: 233–244.
44. Takeuchi, Y. (1986), Global stability in generalized Lotka-Volterra diffusion systems, J. Math. Anal. Appl.,**116**: 209–221.
45. Takeuchi, Y. (1986), Diffusion effect on stability of Lotka-Volterra model, Bull. Math. Biol. **46**: 585–601.
46. Takeuchi, Y. (1989), Cooperative system theory and global stability of diffusion models, Acta Appl. Math. **14**: 49–57.
47. Takeuchi, Y. (1989), Diffusion-mediated persistence in two-species competition Lotka-Volterra model, Math. Biosci. **95**: 65–83.
48. Takeuchi, Y. (1990), Conflict between the need to forage and the need to avoid competition: persistence of two-species model, Math. Biosci. **99**: 181–194.
49. Takeuchi, Y. (1996), *Global Dynamical Properties of Lotka-Volterra Systems*, (World Scientific, Singapore).

50. Takeuchi, Y., J. Cui, R. Miyazaki and Y. Saito (2006), Permanence and periodic solution of dispersal population model with time delays, J.Comp. Appl. Math. **192**: 417–430.
51. Teng, Z. and L. Chen (2003), Permanence and extinction of periodic predator-prey systems in a patchy environment with delay, Non. Anal. R.W.A. **4**: 335–364.
52. Teng, Z. and Z. Lu (2001), The effect of dispersal on single-species nonautonomous dispersal models with delays, J. Math. Biol., **42**: 439–454.
53. Wang, W. and L. Chen (1997), Global stability of a population dispersal in a two-patch environment, Dynamic Systems and Applications, **6**: 207–216.
54. Wang, S., Y. Qu, Z. Jing and Q. Wu (1997), Research on the suitable living environment of the *Rana temporaria chensinensis* larva, Chinese Journal of Zoology, **32(1)**: 38–41.
55. Xu, R. and L. Chen (2001), Persistence and global stability for a delayed nonautonomous predator-prey system without dominating instantaneous negative feedback, J Math. Anal. Appl. **262**: 50–61.
56. Xun, Y. (1990), State disturbance and development of Chinese Giant Pandas, Chinese Wildlife: 9–11.
57. Yange, Y. (1994), Giant panda's moving habit in Poping, Acta Theridogica Sinica **14(1)**: 9–14.
58. Yucun, C. (1994), The urgent problems in reproduction of giant pandas, Chinese Wildlife: 3–5.
59. Zeng, G., L. Chen and J. Chen (1994), Persistence and periodic orbits for two-species nonautonomous diffusion Lotka-Volterra models, Mathl. Comput. Modelling **20**: 69–80.
60. Zhang, J. and L. Chen (1996), Periodic solutions of single-species nonautonomous diffusion models with continuous time delays, Mathl. Comput. Modelling **23**: 17–27.
61. Zhou, Y. (1997), Analysis on decline of wild *Alligator sinensis* population, Sichuan Journal of Zoology **16**: 137–139.

5

Sexual Reproduction Process on One-Dimensional Stochastic Lattice Model

Kazunori Sato

Summary. I consider the stochastic lattice model for sexual reproduction process on one-dimensional lattice investigated by Dickman and Tomé (1991), Noble (1992) and Neuhauser (1994). This model concerns the reproduction to the empty neighboring habitat by a pair of individuals on one-dimensional lattice. Noble (1992) and Neuhauser (1994) mathematically analysed the model with rapid stirring and long-range interaction, respectively. In this chapter, after reviewing the process with rapid stirring briefly, I concentrate on the case with the nearest neighboring interactions and without stirring, and the qualitative features of the dynamics for this model are studied by using pair approximation, which shows the comparative difference from the mean-field approximation.

5.1 Introduction

To date, lattice models in ecology have been successfully applied in explaining various ecological phenomena. From the points of view of mathematics, the interacting particle systems have their own long history (e.g. Liggett 1999 and references therein) based on the basic contact processes introduced by Harris (1974), which can be considered as the SIS epidemic model or logistic model with local competition for space being the limited resource. On the other hand, the procedures or methods of theoretical analyses of lattice models used in statistical physics have been successfully applied to theoretical ecology, and especially, the technique of approximation for the dynamics called pair approximation became popular and neccessary to study both qualitative and quantitative features of lattice models (Matsuda et al. 1992).

In this chapter, I consider the sexual reproduction process on the one-dimensional lattice space, originally proposed by Dickman and Tomé (1991) in the context of autocatalytic reactions, and mathematically studied by Noble (1992) and Neuhauser (1994) for the case of rapid stirring and long-range interaction, respectively. Before their studies, similar processes on two-dimensional lattices were analyzed by several authors (see Chen 1992, Durrett 1999, 1986; Durrett and Neuhauser 1994).

As Noble pointed out, the model with rapid stirring has remarkable characteristics such as dynamical behavior independent of the initial configuration, which can be contrasted to the complete mixing model (or mean-field dynamics) with bistability. I study the dynamics of pair approximation (and triplet decoupling approximation) for this model, compare this to Monte Carlo simulation, and provide qualitative insight for the case without stirring.

5.2 Sexual reproduction process with stirring

Noble (1992) considers the sexual reproduction process with stirring as follows. It seems to be hard to find the correspondence to the biological system in the real world, but one may interpret this model as the abstract and simplest model for the mechanism of self-incompatibility.

I consider $_\varepsilon \xi_t$ as the interacting particle system on $\varepsilon \mathbf{Z}$ or one-dimensional (infinite size of) lattice model defined as the state of the process at time t and the unit of the spatial scale with ε. If $_\varepsilon \xi_t^+$ indicates the process starting from all the sites occupied by "+", then I can be confident of the existence of the equilibrium because of the attractiveness of this process defined below. Also, $_\varepsilon \xi_t(x)$ has either the value "+" or "0", which indicates the site x at time t is either an occupied site (by an individual or organism) or an empty site, respectively. Each individual dies at a constant rate. Each empty site can be given birth by an individual when the adjacent pair of sites is occupied by two individuals at a rate proportinal to the number of adjacent pair of the occupied sites. Choosing the proper time scale, I can use the death rate as 1 and the birth rate as $b/2$ for each pair of occupied sites.

I use the following notation as the probability measure:

$$\rho_{\sigma_{-m},\ldots,\sigma_{-1},\underline{\sigma_0},\sigma_1,\ldots,\sigma_n}(t) = P\{_\varepsilon \xi_t^+(x+k\varepsilon) = \sigma_k \text{ for } -m \leq k \leq n\},$$

where $\sigma_i \in \{0, +\}$ and the underline indicates the focal site. I also define the critical value of the birth rate for the survival as

$$b_c(\varepsilon) = \inf\{b : {}_\varepsilon \xi_t^+ \text{ survives}\}.$$

The dynamics of the model by probability measure for a single site can be written as follows:

$$\begin{aligned}\frac{d}{dt}\rho_\pm &= -\rho_\pm + \frac{b}{2}\rho_{++\underline{0}} + \frac{b}{2}\rho_{\underline{0}++} \\ &\quad + \frac{1}{2}\varepsilon^{-2}(\rho_{\underline{0}+} - \rho_{0\pm} + \rho_{+\underline{0}} - \rho_{\pm 0}) \\ &= -\rho_\pm + \frac{b}{2}\rho_{++\underline{0}} + \frac{b}{2}\rho_{\underline{0}++}.\end{aligned}$$

By the symmetry of the transition rules, the relation $\rho_{++\underline{0}} = \rho_{\underline{0}++}$ holds, and I omit the underline because of the translation invariance of the process. Then, I can obtain the following dynamics:

$$\frac{d\rho_+}{dt} = -\rho_+ + b\rho_{++0}. \tag{1}$$

The first term corresponds to the death of an individual, and the second to the birth by contact of two individuals.

5.2.1 Mean-field approximation

I review the result by mean-field approximation as the first approximation for the present stochastic spatial model (Noble 1992). In this approximation, the interactions occur between randomly chosen sites. so this assumption turns out to be without spatial configuration of the process, i.e. the events are independent of spatial configurations:

$$\rho_{\sigma\sigma'\sigma''} \simeq \rho_\sigma \rho_{\sigma'} \rho_{\sigma''},$$

then Eq. (1) becomes

$$\begin{aligned}\frac{d\rho_+}{dt} &= -\rho_+ + b\rho_+^2 \rho_0 \\ &= -\rho_+ + b\rho_+^2(1-\rho_+),\end{aligned} \tag{2}$$

where I use $\rho_0 = 1 - \rho_+$ in the last equality.

Although I can obtain the solution of Eq. (2) as a rather complicated form of implicit function using logarithmic function and arc tangent, I consider only the ω-limit of the solution. When $b < 4$, there exists only one fixed point: 0. When $b = 4$, there exist two fixed points: $0, \frac{1}{2}$. When $b > 4$ there exist three fixed points: $0, \rho_c = \frac{1-\sqrt{1-4/b}}{2}, \rho_f = \frac{1+\sqrt{1-4/b}}{2}$. The stability analyses reveal the following results. Suppose that $\rho_+(0) > 0$. When $b < 4$, $\lim_{t\to\infty} \rho_+(t) = 0$. When $b = 4$, $\lim_{t\to\infty} \rho_+(t) = \frac{1}{2}$. When $b > 4$,

$$\lim_{t\to\infty} \rho_+(t) = \begin{cases} \rho_f & \text{if } \rho_+(0) > \rho_c, \\ \rho_c & \text{if } \rho_+(0) = \rho_c, \\ 0 & \text{if } \rho_+(0) < \rho_c. \end{cases}$$

This result shows that the mean-field approximation fails to explain the correct dynamics because Noble (1992) proved the independence of the initial configuration with any positive density for the ω-limit of the solution. The mean-field approximation also indicates the discontinuity of the solution in the parameter b at $b_c = 4$.

5.2.2 Pair approximation

Similarly to Eq. (1), I can derive the dynamics of probability measure for the pair of (++) as follows:

$$\frac{d\rho_{++}}{dt} = -2\rho_{++} + b\rho_{++0} + b\rho_{++0+} - \frac{1}{\varepsilon^2}\rho_{++0} + \frac{1}{\varepsilon^2}\rho_{+0+} . \quad (3)$$

I introduce the quantity q as the conditional probability (Matsuda et al. 1992). For example, $q_{\sigma/\sigma'\sigma''}$ corresponds to the probability that the focal site has the state σ with the nearest neighboring site of the state σ' and the next-nearest neighboring site of the state σ''. Similarly, I use $q_{\sigma/\sigma'}$ as the conditional probability that the focal site has the state σ with the nearest neighboring site of the state σ'.

Then, I can construct the dynamics by pair approximation, assuming that

$$\rho_{\sigma\sigma'\sigma''} = \rho_{\sigma\sigma'}q_{\sigma''/\sigma'\sigma} \simeq \rho_{\sigma\sigma'}q_{\sigma''/\sigma'}$$

and

$$\rho_{\sigma\sigma'\sigma''\sigma'''} = \rho_{\sigma\sigma'\sigma''}q_{\sigma'''/\sigma''\sigma'\sigma} \simeq \rho_{\sigma\sigma'}q_{\sigma''/\sigma'}q_{\sigma'''/\sigma''} ,$$

which indicates the interactions with next-nearest neighboring sites are less important than those with the nearest neighboring sites. Besides that, in order to close the system by two variables, ρ_+ and $q_{+/+}$, the following relations are used:

$$\rho_{++} = \rho_+ q_{+/+},$$
$$q_{0/+} = 1 - q_{+/+},$$
$$q_{+/0} = \frac{\rho_+(1 - q_{+/+})}{1 - \rho_+} .$$

Then, the system with (1) and (3) can be approximated as

$$\begin{cases} \dfrac{d\rho_+}{dt} = -\rho_+ + b\rho_+ q_{+/+}(1 - q_{+/+}), \\ \dfrac{d\rho_{++}}{dt} = -2\rho_+ q_{+/+} + b\rho_+ q_{+/+}(1 - q_{+/+}) \\ \quad + b\rho_+ q_{+/+}(1 - q_{+/+})\dfrac{\rho_+(1 - q_{+/+})}{1 - \rho_+} \\ \quad - \dfrac{1}{\varepsilon^2}\rho_+ q_{+/+}(1 - q_{+/+}) + \dfrac{1}{\varepsilon^2}\rho_+(1 - q_{+/+})\dfrac{\rho_+(1 - q_{+/+})}{1 - \rho_+} . \end{cases} \quad (4)$$

Noting that

$$\frac{dq_{+/+}}{dt} = \frac{d(\rho_{++}/\rho_+)}{dt} = -\frac{\rho_{++}}{\rho_+^2}\frac{d\rho_+}{dt} + \frac{1}{\rho_+}\frac{d\rho_{++}}{dt}$$

(Harada and Iwasa 1994), and putting $x = \rho_+, y = q_{+/+}$ into Eq. (4) yields

$$\begin{cases} \dfrac{dx}{dt} = -x + bxy(1 - y), \\ \dfrac{dy}{dt} = -y + by(1 - y)^2 + \dfrac{bxy(1 - y)^2}{1 - x} - \dfrac{1}{\varepsilon^2}y(1 - y) + \dfrac{1}{\varepsilon^2}\dfrac{x}{1 - x}(1 - y)^2 . \end{cases} \quad (5)$$

Within the limit of rapid stirring with $\varepsilon \to 0$ the system (5) gives the steady state as the solution of the following algebraic equations:
$$\begin{cases} -x + bxy(1-y) = 0, \\ -y + \frac{x}{1-x}(1-y) = 0, \end{cases}$$
and it becomes $x = y = 0$, $x = y = \rho_c$ or $x = y = \rho_f$, where ρ_c and ρ_f are the equilibria by mean-field approximation. Therefore, the equilibria by pair approximation coincide with those by mean-field approximation in the case of $\varepsilon \to 0$.

5.3 Sexual reproduction process without stirring

Next, I consider the case without stirring i.e. $\varepsilon \to \infty$. First, the reader should note that the assumption of the mean-field approximation does not account for spatial configuration, so it makes no difference whether the mean-field model includes the stirring or not.

5.3.1 Pair approximation

In the case without stirring, pair approximation (5) reduces as follows:
$$\begin{cases} \dfrac{dx}{dt} = x[by(1-y) - 1], \\ \dfrac{dy}{dt} = \dfrac{y[b(1-y)^2 - 1 + x]}{1-x} \end{cases} \quad (6)$$

Here, I should note that the two-dimensional system by pair approximation is restricted to the region
$$\Omega = \left\{ (y, x) \,\Big|\, 0 \le y \le 1,\ 0 \le x \le \frac{1}{2-y} \right\} \setminus \{(1,1)\},$$
because the two variables y and x correspond to probabilities and then $0 \le \rho_{00} = 1 - \rho_{++} - 2\rho_{0+} = 1 - \rho_+(2 - q_{+/+}) = 1 - x(2-y)$. In addition, I exclude only one point $(1,1)$ on which the dynamics cannot be defined in pair approximation. I can show that the region Ω is positive invariant (Appendix A). When I consider the initial condition restricted to the positive quadrant, it is also convenient to define the following notation:
$$\Omega_+ = \left\{ (y, x) \,\Big|\, 0 < y \le 1,\ 0 < x \le \frac{1}{2-y} \right\} \setminus \{(1,1)\}.$$

This system (6) has the possibility of three equilibria,

$E_{00}: (y_0, x_0) = (0, 0)$,

$E_{+0}: (y_1, x_1) = \left(1 - \dfrac{1}{\sqrt{b}}, 0\right)$,

$E_{++}: (y_2, x_2) = \left(\dfrac{1 + \sqrt{1-4/b}}{2},\ \dfrac{\sqrt{1-4/b}(1 - \sqrt{1-4/b})}{2}\right)$.

I have the trivial equilibrium E_{00} for all $b > 0$. When $0 < b \le 1$, then E_{00} is the unique equilibrium and it becomes globally stable with respect to Ω. When $1 < b < 4$, then E_{+0} also becomes the equilibrium. In this case, E_{00} is an unstable saddle point and E_{+0} is the locally stable boundary equilibrium, and then I know that E_{+0} is globally stable with respect to $\Omega_+ \cup \{(y,0) \mid 0 < y \le 1\}$. When $b \ge 4$, then the internal equilibrium E_{++} also exists and in this case I can conclude that E_{++} is globally stable with respect to Ω_+ (Appendix B, Fig. 5.2).

5.3.2 Triplet decoupling approximation

I can construct the dynamics by triplet of contact sites assuming that

$$\rho_{\sigma\sigma'\sigma''\sigma'''} = \rho_{\sigma\sigma'\sigma''} q_{\sigma'''/\sigma''\sigma'\sigma} \simeq \rho_{\sigma\sigma'\sigma''} q_{\sigma'''/\sigma''\sigma'}$$

and

$$\rho_{\sigma\sigma'\sigma''\sigma'''\sigma''''} = \rho_{\sigma\sigma'\sigma''\sigma'''} q_{\sigma''''/\sigma'''\sigma''\sigma'\sigma} \simeq \rho_{\sigma\sigma'\sigma''} q_{\sigma'''/\sigma''\sigma'} q_{\sigma''''/\sigma'''\sigma''} \, ,$$

which indicates the interactions with distant neighboring sites are less important than those with the closer neighboring sites (see also Sect. 5.2). Using the relationship between ρ and q, I can obtain the system of triplet decoupling approximation in the following:

$$\begin{cases} \dfrac{d\rho_+}{dt} = -\rho_+ + b\rho_{++0} \, , \\[4pt] \dfrac{d\rho_{++}}{dt} = -2\rho_{++} + b\rho_{++0} + b\rho_{++0+} \\[4pt] \qquad \simeq -2\rho_{++} + b\rho_{++0} + b\rho_{++0} \dfrac{\rho_+ - \rho_{++} - \rho_{+00}}{\rho_+ - \rho_{++}} \, , \\[4pt] \dfrac{d\rho_{+++}}{dt} = -3\rho_{+++} + b\rho_{++0} + b\rho_{++0++} + b\rho_{+0++} \\[4pt] \qquad \simeq -3\rho_{+++} + b(\rho_{++} - \rho_{+++}) \\[4pt] \qquad + b(\rho_{++} - \rho_{+++}) \dfrac{\rho_+ - \rho_{++} - \rho_{+00}}{\rho_+ - \rho_{++}} \dfrac{\rho_{++} - \rho_{+++}}{\rho_+ - \rho_{++}} + b\rho_{++0} \dfrac{\rho_+ - \rho_{++} - \rho_{+00}}{\rho_+ - \rho_{++}} \, , \\[4pt] \dfrac{d\rho_{+00}}{dt} = -\rho_{+00} - \dfrac{b}{2}\rho_{++00} - \dfrac{b}{2}\rho_{+00++} + \dfrac{b}{2}\rho_{++000} + \rho_{++0} + \rho_{+0+} \\[4pt] \qquad \simeq -\rho_{+00} - \dfrac{b}{2}(\rho_{++} - \rho_{+++})(1 - \dfrac{\rho_+ - \rho_{++} - \rho_{+00}}{\rho_+ - \rho_{++}}) \\[4pt] \qquad - \dfrac{b}{2}(\rho_{++} - \rho_{+++})(1 - \dfrac{\rho_+ - \rho_{++} - \rho_{+00}}{\rho_+ - \rho_{++}}) \dfrac{\rho_{+00}}{1 - 2\rho_+ + \rho_{++}} \\[4pt] \qquad + \dfrac{b}{2}(\rho_{++} - \rho_{+++})(1 - \dfrac{\rho_+ - \rho_{++} - \rho_{+00}}{\rho_+ - \rho_{++}})(1 - \dfrac{\rho_{+00}}{1 - 2\rho_+ + \rho_{++}}) \\[4pt] \qquad + (\rho_{++} - \rho_{+++}) + (\rho_+ - \rho_{++} - \rho_{+00}) \, . \end{cases}$$

This system is closed by the four variables ρ_+, ρ_{++}, ρ_{+++} and ρ_{+00}. The numerical calculation suggests that b_c, the critical value for survival, is about 5.89.

5.3.3 Monte Carlo simulation

I execute the Monte Carlo simulation to estimate the steady states of the model. The total size of the lattice (i.e., the number of the total sites) is 10000, and the periodic boundary condition (i.e., the leftmost site is connected to the rightmost site) is used to avoid the effect of the boundary. I terminate the calculation at 10000 MCS (= Monte Carlo steps), and show the average for the last 100 MCS, along with the equilibrium values by mean-field approximation, pair approximation, and triplet decoupling approximation in Fig. 5.1. These graphs indicate that the equlibrium value ρ_+ suddenly but continuously increases for the birth rate b, which is conjectured by Durrett and Neuhauser (1994), but mean-field approximation fails to explain this. On the other hand, pair approximation succeeds in providung not only this continuity but also closer estimation (and triplet decoupling approximation gives better quantitative results).

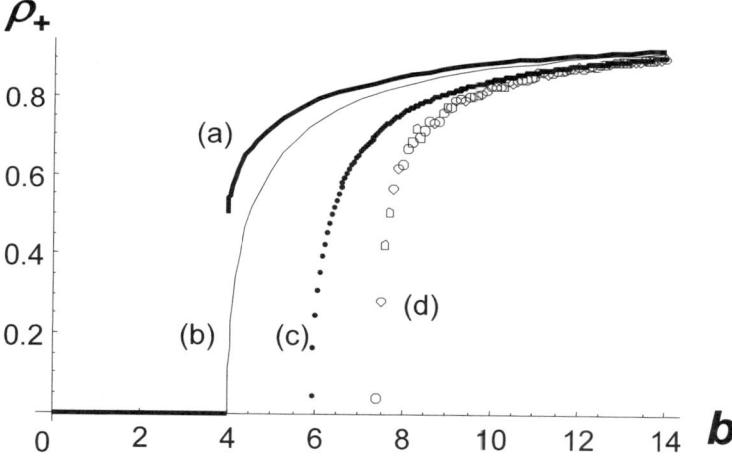

Fig. 5.1. Equilibrium values ρ_+ as a function of b for the model without stirring. (a) mean-field approximation, (b) pair approximation, (c) triplet decoupling approximation, (d) Monte Carlo simulation

5.4 Discussion

Noble (1992) rigorously analysed the sexual reproduction process with rapid stirring, and he showed that the dynamical behavior toward the unique equilibrium can be compared to the corresponding PDE model. By contrast, mean-field dynamics, or PDE without spatial heterogeneity in the initial configuration, gives the bistability dependent on the initial state. The critical birth rate b_c, which is larger than the value given by mean-field dynamics, can be obtained by the traveling wave solution of the corresponding PDE. It seems to be rather difficult to intuitively understand the spatial heterogeneity causes that difference, but this may be better understood when I analyse the dynamics of the spatial pattern.

Pair approximation seems to have the transition from unique stability to bistability at some stirring rate ε_c. For smaller stirring rates, some spatial characteristics such as stronger spatial clumping or higher spatial correlation probably play an important role in producing the difference vis-à-vis mean-field dynamics. The critical point by pair approximation is the same as that by mean-field, but the triplet decoupling approximation suggests to give a larger value, which indicates a quantitative improvement. Besides that, as Durrett and Neuhauser (1994) conjectured and in contrast to Dickman and Tomé (1991), I also expect the continuity of the equilibria on the parameter b obtained by pair approximation.

As a future problem, I aim to study the present model with a finite stirring rate. I expect that this model gives the unique stable equilibrium and larger critical point for survival than that obtained for the model with rapid stirring, as I show the case without stirring by pair approximation in this chapter.

Acknowledgement. I sincerely thank Prof. Takeuchi for his helpful comments on the earlier version of the draft.

Appendix A: Positive invariance of Ω

Here I set two axes of y and x as the horizontal and the vertical, respectively, for ease of handling (Fig. 5.2). In this Appendix A, I concentrate on the case of $b > 4$ (Fig. 5.2d), but I can check the positive invariance of Ω for other cases in a similar way.

To examine the positive invariance, it is enough to consider the behavior of the solutions on the boundaries. On the x-axis, the solution moves toward the origin. On the other hand, on the y-axis, the solution moves toward the point $(\hat{y}, 0)$, starting from the point in the interval $(0,1]$ on y-axis, where $(\hat{y}, 0)$ is the unique intersection between the null cline of $\frac{dy}{dt} = 0$ and the y-axis, and putting $\hat{y} = 1 - \frac{1}{\sqrt{b}}$. On the line $y = 1$, $\frac{dy}{dt} = -1 < 0$ implies that the solutions move inside Ω.

Next, I show that the solutions do not go outside Ω by evaluating the magnitude of the gradients at each point on the boundary $x = h(y) = \frac{1}{2-y}$ for $y \in I_1 = (y_-, y_+)$ and $y \in I_2 = (y_*, 1]$. Here, $(y_\pm, 0)$ are the intersections between the null cline of $\frac{dx}{dt} = 0$ and the y-axis, and I put $y_\pm = \frac{1 \pm \sqrt{1-4/b}}{2}$, obtained from the quadratic equation $f(y) = -1 + by(1-y) = 0$. Besides y_*

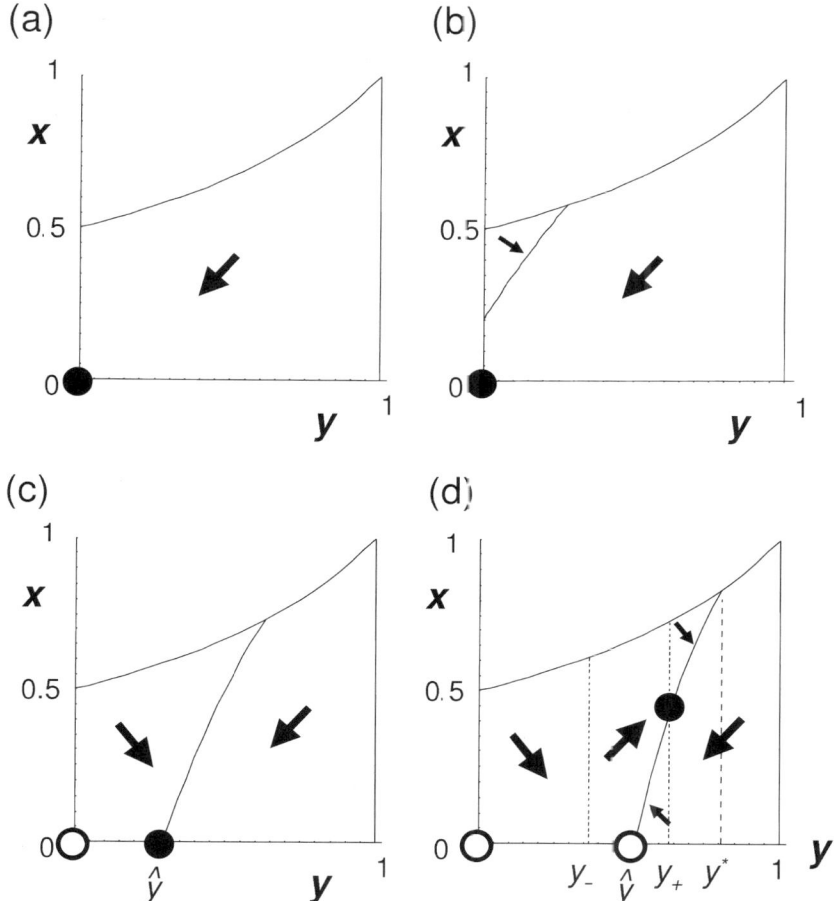

Fig. 5.2. Phase portrait of pair approximation; $x = p_+, y = q_{+/+}$. The graphs change depending on the values of b: **a** $0 < b \le 1/2$, **b** $1/2 < b \le 1$, **c** $1 < b \le 4$, **d** $b > 4$. The *arrow* represents the flow of the points in each region. The *filled* and *empty circles* correspond to the stable and unstable equilibrium, respectively

and 1 are the y-values in the intersections between the null cline of $\frac{dy}{dt} = 0$ other than the y-axis and the boundary $x = h(y)$, and I put $y_* = \frac{3-\sqrt{1+4/b}}{2}$, which is the solution of the quadratic equation $g(y) = b(1-y)(2-y) - 1 = 0$ within $[0, 1]$. The magnitude of the gradient on the point (y, x) on the boundary $x = h(y)$ can be evaluated by $\frac{dh(y)}{dy}$, and the slope of the solution by $\frac{\dot{x}}{\dot{y}}$. Therefore, if I can show $\frac{dh(y)}{dy} - \frac{\dot{x}}{\dot{y}} > 0$ in the case $y \in I_1$ and $\frac{dh(y)}{dy} - \frac{\dot{x}}{\dot{y}} < 0$ in the case $y \in I_2$, then I can conclude that Ω is positive invariant.

In the former case, $y \in I_1$, I have $f(y) > 0$ because y_- and y_+ are the solutions of the quadratic equation $f(y) = 0$. Then, I have

$$\frac{dh(y)}{dy} - \frac{\dot{x}}{\dot{y}} = \frac{1}{(2-y)^2} - \frac{x[-1+by(1-y)]}{y\left[-1+b(1-y)^2+\frac{bx(1-y)^2}{1-x}\right]}$$

$$= \frac{1}{y(2-y)^2} \frac{2(1-y)}{[by(1-y)-1]+2b(1-y)^2} > 0.$$

In the latter case, $y \in I_2$, note that $g(y) < 0$ because y_* is the solution of the quadratic equation $g(y) = 0$. Then, I have

$$\frac{dh(y)}{dy} - \frac{\dot{x}}{\dot{y}} = \frac{1}{(2-y)^2} - \frac{x[-1+by(1-y)]}{y\left[-1+b(1-y)^2+\frac{bx(1-y)^2}{1-x}\right]}$$

$$= \frac{1}{y(2-y)^2} \frac{2(1-y)}{b(1-y)(2-y)-1} < 0.$$

Appendix B: Global stability analysis of internal equilibrium by pair approximation

I can show the global stability of the internal equilibrium by using the Butler–McGehee theorem, Poincaré–Bendixson theorem and Dulac criterion for the two-dimensional autonomous pair approximation system (Smith and Waltman 1995). For a start, the boundedness of the solutions is clear from the positive invariance of Ω by Appendix A. Next, I show that any trajectory starting from Ω_+ cannot contain the origin and the boundary equilibrium in its ω limit set by Butler–McGehee theorem. Poincaré-Bendixson theorem tells us that the ω limit set of any such trajectory must be either a periodic orbit or an internal equilibrium. Finally, I can deny the possibility of the ω limit set being a periodic orbit by finding a Dulac function.

(i) Local stability of three equiliria E_{00}, E_{+0} and E_{++}

I obtain the Jacobian matrix for equilibrium $(\bar{y}, \bar{x})$ as follows:

$$\begin{pmatrix} -\frac{-(1-x)+b(1-y)(1-3y)}{1-\bar{x}} & \frac{b\bar{y}(1-\bar{y})^2}{(1-\bar{x})^2} \\ b\bar{x}(1-2\bar{y}) & b\bar{y}(1-\bar{y})-1 \end{pmatrix}.$$

This matrix becomes the following in the case of E_{00}:

$$\begin{pmatrix} b-1 & 0 \\ 0 & -1 \end{pmatrix},$$

then it indicates that E_{00} is a saddle point. Indeed, as I mentioned above (Appendix A), an interval $[0, 1/2]$ on the x-axis and an interval $(0, y^*)$ on the y-axis correspond to a stable and an unstable manifold for E_{00}, respectively. Therefore, starting from Ω_+, E_{00} itself cannot be the entire ω limit set. By using the Butler–McGehee theorem as well as the fact that our ω limit set is closed and invariant, if the ω limit set contains E_{00}, then it must also contain an interval $[0, 1/2]$ on the x-axis. However, the ω limit set cannot contain a point $(0, 1/2)$ on the x-axis, which is not approached from any direction, and so there is no possibility of E_{00} being in the ω limit set. Next, I consider the case of E_{+0} with the Jacobian matrix as

$$\begin{pmatrix} -2(\sqrt{b}-1) & 1-\frac{1}{\sqrt{b}} \\ 0 & \sqrt{b}-2 \end{pmatrix},$$

which tells us that E_{+0} is also a saddle point. Indeed (Appendix A), an interval $[0, 1]$ on y-axis becomes a stable manifold for E_{+0}. An eigenvector $\left(1, \frac{\sqrt{b}(3-4/\sqrt{b})}{1-1/\sqrt{b}}\right)$ for a positive eigenvalue $\sqrt{b}-2$ points into Ω_+. Similarly as for the case of E_{00}, starting from Ω_+, E_{+0} itself cannot equal the ω limit set. Noting that our ω limit set is closed and invariant, the Butler–McGehee theorem tells us that if the ω limit set contains E_{+0}, then it must also contain an interval $[0, 1]$ on the y-axis. However, as mentioned above, E_{00} cannot be contained in the ω limit set, which represents a contradiction. Then, I can conclude that E_{+0} is not contained in the ω limit set. For the internal equilibrium E_{++}, substituting the equilibrium values

$$y_2 = \frac{1+\sqrt{1-4/b}}{2} \quad \text{and} \quad x_2 = \frac{b\sqrt{1-4/b}(1-\sqrt{1-4/b})}{2}$$

into the Jacobian matrix results in the following:

$$\begin{pmatrix} \frac{4}{b[\sqrt{1-4/b}-(1-2/b)]} & \frac{2(1-\sqrt{1-4/b})}{b^2[(1-2/b)-\sqrt{1-4/b}]^2} \\ -\frac{b(b-4)(1-\sqrt{1-4/b})}{2} & 0 \end{pmatrix} = \begin{pmatrix} - & + \\ - & 0 \end{pmatrix},$$

where the last matrix represents the signature of each element, and then I can find that the real parts of all the eigenvalues of this matrix are negative.

(ii) No cyclic solution by Dulac function $\beta(x, y)$

When I choose $\beta(x, y) = \frac{1}{(1-x)y}$, then

$$\frac{\partial}{\partial x}\left[\beta(x,y)\frac{dx}{dt}\right] + \frac{\partial}{\partial y}\left[\beta(x,y)\frac{dy}{dt}\right]$$

$$= \frac{\partial}{\partial x}\left[\frac{-x+bxy(1-y)}{(1-x)y}\right] + \frac{\partial}{\partial y}\left[\frac{-y+by(1-y)^2 + \frac{bxy(1-y)^2}{1-x}}{(1-x)y}\right]$$

$$= -\frac{by(1-y)+1}{y(1-x)^2} < 0$$

for all x and y, which indicates the absence of cyclic solution.

References

1. Chen, H.N. (1992), On the stability of a population growth model with sexual reproduction on Z^2, Ann. Probab. **20**, 232–285
2. Dickman, R. and T. Tomé (1991), First-order phase transition in a one-dimensional nonequilibrium model, Phys. Rev. A **44**, 4833–4838
3. Durrett, R. (1999), Stochastic spatial models, SIAM Review **41**, 677–718
4. Durrett, R. (1986), Some peculiar properties of a particle system with sexual reproduction. In: *Lecture Note in Mathematics*, vol 1212, ed by P. Tăutu (Springer, New York) pp 106–111
5. Durrett, R. and C. Neuhauser (1994), Particle systems and reaction-diffusion equations, Ann. Probab. **22**, 289–333
6. Harada, Y. and Y. Iwasa (1994), Lattice population dynamics for plants with dispersing seeds and vegetative propagation, Res. Popul. Ecol. **36**, 237–249
7. Harris, T.E. (1974), Contact interactions on a lattice, Ann. Probab. **2**, 969–988
8. Liggett, T.M. (1999), *Stochastic Interacting Systems: Contact, Voter and Exclusion Processes*, (Springer, Berlin Heidelberg)
9. Matsuda, H. et al. (1992), Statistical mechanics of population – The Lotka–Volterra modell –, Prog. Theor. Phys. **88**, 1035–1049
10. Neuhauser, C. (1994), A long range sexual reproduction process, Stoch. Proc. Appl. **53**, 193–220
11. Noble, C. (1992), Equilibrium behavior of the sexual reproduction process with rapid diffusion, Ann. Probab. **20**, 724–745
12. Smith, H. L. and P. Waltman (1995), *The Theory of the Chemostat*, (Cambridge University Press, Cambridge)

6

A Mathematical Model of Gene Transfer in a Biofilm

Mudassar Imran and Hal L. Smith

Summary. Based on our previous work, a model of plasmid transfer between micro-organisms in a heterogeneous environment consisting of a biofilm immersed in a fluid medium is constructed. A review of previous modeling of gene transfer is provided in order to place our work in context. The key question is whether the plasmid can persist in the bacterial population. We answer this question by constructing a basic reproductive number which takes into account the advantages conferred by the plasmid and its costs to the bacterial host.

6.1 Introduction

Plasmids, small circular strands of DNA separate from the main genome of the organism, are common in natural bacterial populations such as soils, lakes and stream and in the gut of mammals. They often carry genes for such beneficial factors as resistance to antibiotics and heavy metals, the ability to ferment sugars, or to produce toxins. Some carry genes for pili production and mating pair formation that allow the infectious transfer of the plasmid to other bacteria – a process called conjugation. However, many plasmids have no known function in bacteria and may simply be parasitic. Vertical transmission of plasmids occurs during cell division when the plasmids in the cell are duplicated and partitioned among the daughter cells; rarely, however, one daughter cell may end up without plasmid while the other daughter cell receives multiple copies. This loss of plasmid is referred to a segregative loss. Furthermore, there is some cost to an organism carrying plasmids since the cell may produce plasmid gene products and must duplicate it during cell division which leads to a reduced reproductive rate. See Simonsen (1991) for a readable review, particularly of modeling aspects, and Summers (1996) for a general review. Because the benefits of plasmid carriage, if any, depend on ever-changing environmental conditions while the costs are always present, a longstanding focus of theoretical studies has been to determine conditions under which plasmids can be maintained in bacterial populations. See Stewart and Levin (1977), Levin and Rice (1980) and Bergstrom, Lipsitch, and Levin (2000) for modeling results related to this problem.

According to Angles and Goodman (2000):

> Biofilms are environments of high microbial density where cell-cell contact is likely. Such conditions create a favorable niche for the spread of self-transmissible as well as mobilisable plasmids among members of the bacterial communities. Studies have demonstrated plasmid transfer among bacteria in a wide range of biofilm habitats, including the surface of stones in a river, the air-water interface, surfaces in soil and water microcosms, plant surfaces and insect as well as animal intestinal surfaces.

In a recent paper Ghigo (Ghigo 2001) established that several natural conjugative plasmids express factors that induce some bacteria to form biofilms. Experimental studies showed that a strain of *E. Coli* bearing a certain plasmid formed a thick biofilm within one day while those not carrying the plasmid produced no macroscopically observable biofilm. Interestingly, Ghigo's results suggest that the pili responsible for the horizontal transfer of the plasmid, may also act as an adhesion factor for cell-to-surface contact. See also Pratt and Kolter (1998) and O'Toole and Kolter (1998). Ghigo points out the many beneficial aspects for bacteria in biofilms relative to the fluid environment and speculates that such factors "may provide a rationale for the unexplained vertical maintenance of the numerous uninfectious cryptic plasmids found in natural populations". He also observes that by inducing bacteria to form the denser communities characteristic of biofilms the plasmid increases the likelihood of its own horizontal transfer via conjugation.

In this chapter, we explore the suggested link between plasmid maintenance and biofilms by modifying slightly the mathematical model proposed by us in (Imran et al. preprint) of a bacterial population consisting of plasmid-bearing and plasmid-free organisms in a continuous culture with a surface on which a biofilm may form. The question we address is under what circumstances can the plasmid be maintained in a population. Heuristically, the advantageous genes carried by the plasmid together with the ability of the plasmid-bearing organism to pass the plasmid to other organisms must compensate for the energetic cost of bearing the plasmid and the occasional segregative loss of the plasmid during cell division. We seek to quantify this trade-off. Our model builds on the plasmid model of Stephanopoulus and Lapidus (1988) and Ryder and DiBiasio (1984), includes conjugation terms used by Stewart and Levin (1977), and models the biofilm following the model of Pilyugin and Waltman (1999). Consequently, we briefly review these models in order that the foundation of our model is made more clear.

Two cases are considered: (1) the plasmid is parasitic, conferring no advantage on its host, and (2) the plasmid codes for enhanced biofilm forming ability in its bacterial host which in its absence can form only a macroscopically unobservable biofilm. In the first case, the question is under what circumstances can a parasitic plasmid can be maintained. In the second case, the question is under what circumstances can the ability to form a robust biofilm community in which conjugative transfer of the plasmid may occur be sufficiently advantageous for the plasmid-bearing organism to compensate for the energetic cost of bearing the plasmid and the segregative loss of the plasmid. In each case, we provide a quantitative expression of a potential mechanism which may be significant in plasmid maintenance.

Our work corroborates the conjectures of Ghigo. The availability of colonalizable surfaces that provide a selective advantage for an organism carrying a plasmid containing a biofilm-enhancing gene may contribute to the maintenance of such plasmids in natural bacterial populations.

The same models developed in this paper could also be used to study the important phenomena of horizontal spread of antibiotic resistance in the gut. Rather than assuming the plasmid codes for enhanced biofilm forming ability one would assume that it codes for antibiotic resistance. Selection for the resistant strain could, of course, be arranged by adding antibiotic. Ingestion of bacteria containing plasmid coding for antibiotic resistance could lead to the spread of resistance to the gut microflora. This phenomena may play a significant role in the proliferation of antibiotic resistant pathogens (Summers 1996).

6.2 A model of plasmid transfer with wall growth

We consider a population of bacteria in a continuous culture which colonize both the fluid environment and a portion of a surface immersed in the

Table 6.1. Model parameters for the chemostat: t =time, m = mass, l =length

Symbol	Description	Dimension
u, u_+	biomass concentration of planktonic bacteria.	ml^{-3}
w, w_+	areal biomass density of adherent bacteria.	ml^{-2}
β, β_+	sloughing rate.	t^{-1}
α, α_+	rate constant of adhesion.	t^{-1}
S	concentration of limiting substrate.	ml^{-3}
S^0	concentration of the substrate in the feed.	ml^{-3}
γ	yield constant.	—
a	half saturation constant.	ml^{-3}
m	maximum growth rate of plasmid-free organism.	t^{-1}
c	fractional energetic cost of plasmid carriage, $0 < c < 1$.	—
q	fractional segregation loss factor, $0 < q < 1$	—
D	dilution rate.	t^{-1}
μ	biofilm conjugational transfer parameter.	$l^2(mt)^{-1}$
$\bar{\mu}$	planktonic conjugational transfer parameter.	$l^3(mt)^{-1}$

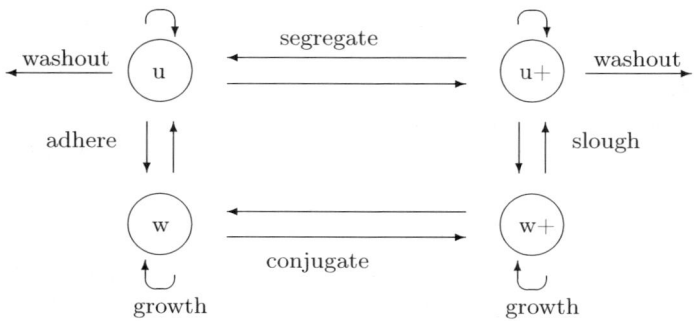

Fig. 6.1. Flow chart of biomass flow between model compartments, u, u_+, w, w_+

fluid. Bacteria are labelled according to their location (fluid or wall: the former called planktonic cells, the latter called adherent cells) and according to whether or not they possess the plasmid of interest (subscript '+' means they have plasmid). Let u (u_+) denote the density of planktonic plasmid-free (plasmid-bearing) organism and w (w_+) denote the areal density of wall-adherent plasmid-free (plasmid-bearing) organism. These populations are supported by the substrate S in continuous culture.

Model parameters are described in the Table 6.1 and a schematic diagram of the model is depicted below it. Bacterial variables and parameters without the "+" sign refer to plasmid-free cells while those with subscript "+" refer to plasmid-bearing cells.

The model equations in the setting of a continuous culture of volume V, colonizable surface area A and flow rate Φ takes the form ($D = \Phi/V$, $\delta = A/V$):

$$\begin{aligned}
S' &= D(S^0 - S) - \gamma^{-1}[f_u(S)u + f_u(S)(1-c)u_+] \\
&\quad - \gamma^{-1}[\,\delta f_w(S)w + \delta f_w(S)(1-c)w_+] \\
u' &= (f_u(S) - D)u + qf_u(S)(1-c)u_+ - \alpha u + \beta \delta w - \bar{\mu}uu_+ \\
w' &= f_w(S)w + qf_w(S)(1-c)w_+ + \alpha \delta^{-1}u - \beta w - \mu ww_+ \\
u'_+ &= [f_u(S)(1-c)(1-q) - D]u_+ - \alpha_+ u_+ + \beta_+ \delta w_+ + \bar{\mu}uu_+ \\
w'_+ &= f_w(S)(1-c)(1-q)w_+ + \alpha_+ \delta^{-1}u_+ - \beta_+ w_+ + \mu ww_+
\end{aligned} \tag{1}$$

These equations represent a modification of the model first proposed in (Imran et al. preprint). See the discussion section for a description of this modification. Because the model is an amalgamation of models in the literature we do not give a detailed description of it here. Instead, we indicate below how it was constructed from earlier models and, in subsequent sections, review these earlier and simpler models. In this way, the basic features of the model may be discussed in a simpler setting without all the distractions present in the full model. After doing this, we return to the full model.

In the absence of the plasmid-bearing organism ($u_+ = w_+ = 0$), the system reduces to the wall growth model of Pilyugin and Waltman(1999):

$$\begin{aligned}
S' &= D(S^0 - S) - \gamma^{-1}\left[f_u(S)u + \delta f_w(S)w\right] \\
u' &= (f_u(S) - D)u - \alpha u + \beta \delta w \\
w' &= f_w(S)w + \alpha \delta^{-1} u - \beta w
\end{aligned} \qquad (2)$$

This system is described in detail in the next section.

Neglecting the wall-attached population ($w = w_+ = 0$ and $\alpha = 0$) in (1) we obtain a model of plasmid transfer in the fluid environment of the chemostat:

$$\begin{aligned}
S' &= D(S^0 - S) - \gamma^{-1}[f_u(S)u + f_{u_+}(S)u_+] \\
u' &= (f_u(S) - D)u + q f_{u_+}(S)u_+ - \bar{\mu} u u_+ \\
u'_+ &= [f_{u_+}(S)(1-q) - D]u_+ + \bar{\mu} u u_+ \\
f_{u_+}(S) &:= (1-c)f_u(S)
\end{aligned} \qquad (3)$$

This model is similar to the classic model of Stewart and Levin (1977). We consider and compare both of these in Sect. 6.4.2.

Ignoring the wall population and plasmid transfer (conjugation), but not segregation, the system reduces to a special case of the model of Stephanopoulus and Lapidus (1988) for (non-infectious) plasmid-bearing organisms in the chemostat:

$$\begin{aligned}
S' &= D(S^0 - S) - \gamma^{-1}[f_u(S)u + f_{u_+}(S)u_+] \\
u' &= (f_u(S) - D)u + q f_{u_+}(S)u_+ \\
u'_+ &= [f_{u_+}(S)(1-q) - D]u_+ \\
f_{u_+}(S) &:= (1-c)f_u(S)
\end{aligned} \qquad (4)$$

Stephanopoulus and Lapidus borrowed ideas from the earlier work of Ryder and DiBiasio (1984). System (4) and similar models are especially relevant to issues in biotechnology involving the production of biologically useful compounds by genetically altered organisms. See Hsu et al. (1994, 1997, 2004).

As (1) is inspired by these earlier models, it is also perhaps best understood once one is familiar with them. We review them in the sections immediately following before returning to analyze (1).

6.3 Pilyugin–Waltman model

Pilyugin–Waltman (1999) proposed a simple chemostat model with wall growth in the form of three nonlinear differential equations. The key difference between their model and the standard chemostat model (see e.g. Smith and Waltman 1995) is that the population growing on the wall does not wash out of the chemostat. Due to this modification the basic conservation principle of the chemostat is lost so the system is no longer reducible to a planar system. With some change in notation, their model is given by

$$S' = D(S^0 - S) - \gamma^{-1} f_u(S) u - \gamma^{-1} \delta f_w(S) w$$
$$u' = (f_u(S) - D) u - \alpha u + \beta \delta w \qquad (5)$$
$$w' = f_w(S) w + \alpha \delta^{-1} u - \beta w$$

where u denotes the volume density of the organism in the fluid (planktonic cells) and w denotes the areal density of the organism on the wall (adherent cells). Planktonic cells adhere to the wall at rate α and adherent cells slough off the wall at rate β. D is the dilution rate, γ is the yield coefficients expressing the proportionality between the uptake rate and growth rate. The nutrient uptake rate $f_u(S)$ should satisfy

$$f_u(0) = 0, \quad f_u'(S) > 0.$$

In practice, they are often taken to be of Michaelis–Menten form

$$f_u(S) = \frac{mS}{a+S}, \quad m, a > 0. \qquad (6)$$

The same conditions hold for f_w, which may be distinct from f_u. Parameter δ is the ratio of the colonizable area A to the volume V of the chemostat. It can be scaled out of the system by replacing w by δw in the model which we routinely do in computing Jacobian matrices below.

A key parameter in the model is the mean residence time (MRT) of a bacterial cell in the chemostat. From the Appendix, we have

$$\text{MRT} = \left[\frac{2}{D + \alpha + \beta - \sqrt{(D + \alpha + \beta)^2 - 4\beta D}} \right] \qquad (7)$$

Because α and β are assumed to be positive, there are only two possible types of steady states, the washout steady state $(S^0, 0, 0)$ and possibly one or more survival steady states having the form $(\bar{S}, \bar{u}, \bar{w})$ with all components positive. The stability of the washout steady state can be determined by the eigenvalues of the variational matrix at $(S^0, 0, 0)$

$$J := \begin{pmatrix} -D & -\gamma^{-1} f_u(S^0) & \gamma^{-1} f_u(S^0) \\ 0 & f_u(S^0) - D - \alpha & \beta \\ 0 & \alpha & f_w(S^0) - \beta \end{pmatrix}$$

We denote by A the lower right two-by-two sub-matrix of J and by $s(A)$ its stability modulus, the maximum of its two real eigenvalues. Evidently, the washout steady state $(S^0, 0, 0)$ is hyperbolically stable if $s(A) < 0$ and unstable if $s(A) > 0$.

The survival steady state can be described most efficiently by introducing the quasi-positive irreducible matrix function of S given by

$$B(S) := \begin{pmatrix} f_u(S) - D - \alpha & \beta \\ \alpha & f_w(S) - \beta \end{pmatrix} \quad (8)$$

In order that $(\bar{S}, \bar{u}, \bar{w})$ be a positive steady state, $(\bar{u}, \delta\bar{w})$ must be a positive eigenvector corresponding to the zero eigenvalue of $B(\bar{S})$. As $S \to B(S)$ is increasing (along the diagonal), Perron–Frobenius theory (Berman and Plemmons 1979) implies that $S \to s(B(S))$ is strictly increasing so there can be at most one value of S at which $s(B(S)) = 0$. Since $A = B(S^0)$ and $s(B(0)) < 0$, we see that if $s(A) > 0$, there is a unique $\bar{S} \in (0, S^0)$ such that $s(B(\bar{S})) = 0$. Then, $(\bar{u}, \delta\bar{w})$ is uniquely determined up to a positive multiple, p, as the positive eigenvector of $B(\bar{S})$. This scalar multiple p is uniquely determined by the steady state equation $S' = 0$ when $\bar{S} < S^0$. If $s(A) \leq 0$, then there may be no $\tilde{S}$ for which $s(B(\bar{S})) = 0$ and even if there is one, $\bar{S} \geq S^0$ so no survival steady state exists.

The main result is the following:

Theorem 6.3.1 [PILYUGIN & WALTMAN] *The following hold for (5):*

(a) *The washout state is globally attracting when it is locally asymptotically stable in the linear approximation, i. e., when $s(A) < 0$.*

(b) *there is a positive "survival" steady state if and only if the washout state is unstable in the linear approximation. When it exists, it is unique and asymptotically stable in the linear approximation.*

(c) *If the washout steady is unstable, then the bacterial population persists. More precisely, there exists $\epsilon > 0$, independent of initial data, such that for all solutions of (5) satisfying $u(0) + \delta w(0) > 0$, there is $T > 0$ such that*

$$u(t) + \delta w(t) > \epsilon, \quad t > T.$$

(d) *If $f_u = f_w$, then the washout state is stable if $R_0 := MRT \cdot f_u(S^0) < 1$ and unstable when $R_0 > 1$. In the latter case, the survival steady state $(\bar{S}, \bar{u}, \bar{w})$ attracts all solutions with $u(0) + \delta w(0) > 0$.*

In part (d), R_0 represents the number of progeny produced by a single cell introduced into the washout steady state.

Pilyugin and Waltman establish the global stability assertion in part (d) by passing to new variables $z = u + \delta w$ and $v = u/z$ and noting that one can reduce the dimension by one since v converges. An interesting open problem is to show that the global stability assertion in (d) holds more generally.

Part (a) is not contained in the results of Pilyugin and Waltman 1999 so we give the argument here. If $s(A) < 0$ then $s(B(S^0 + \epsilon)) < 0$ for sufficiently small $\epsilon > 0$ by continuity of the stability modulus. The first of Eqs. (5) implies that $S' \leq D(S^0 - S)$ so there exists $T > 0$ such that $S(t) < S^0 + \epsilon$ for $t \geq T$.

Consequently, for $t \geq T$,

$$\begin{aligned} u' &\leq (f_u(S^0 + \epsilon) - D)u - \alpha u + \beta \delta w \\ w' &\leq f_w(S^0 + \epsilon)w + \alpha \delta^{-1} u - \beta w \end{aligned} \quad (9)$$

By a well-known comparison theorem (see Theorem B.1 in (Smith and Waltman 1995)), it follows that

$$(u(t), w(t)) \leq (U(t), W(t)), \quad t \geq T$$

where $(U(t), W(t))$ satisfies the linear system obtained by replacing the inequalities by equalities in (9) and the initial conditions $(U(T), W(T)) = (u(T), w(T))$. Because $s(B(S^0 + \epsilon)) < 0$, we conclude that $(U(t), W(t)) \to (0,0)$ as $t \to \infty$ so the same holds for $(u(t), w(t))$, completing the argument.

6.4 Models of plasmid transfer without wall growth

6.4.1 Stewart and Levin model

Stewart and Levin (1977) presented a model that describes the dynamics of conjugationally transmitted plasmids in bacterial populations. They also analyzed the steady state properties of the model. With some change in notation, their model is

$$\begin{aligned} S' &= D(S^0 - S) - \gamma^{-1} f_u(S) u - \gamma_+^{-1} f_{u_+}(S) u_+ \\ u' &= (f_u(S) - D)u + q u_+ - \bar{\mu} u u_+ \\ u'_+ &= [f_{u_+}(S) - D - q] u_+ + \bar{\mu} u u_+ \end{aligned} \quad (10)$$

where u and u_+ are plasmid-free and plasmid-bearing bacterial population densities and S is the concentration of substrate on which they grow. These populations reproduce at rates $f_u(S)$ and $f_{u_+}(S)$ respectively, with properties as in the previous section. Parameters γ^{-1} and γ_+^{-1} are yield coefficients.

Stewart and Levin model conjugation as a mass action type infectious process for the reaction $u + u_+ \to 2u_+$ with infectious rate constant $\bar{\mu}$. This mass action model of conjugation, similar to that used in epidemiological modeling (Diekmann and Heesterbeek 2000), will be used throughout this paper. Segregation is modelled as if a plasmid-bearing cell has per unit time probability q of losing its plasmid and reverting to a plasmid-free organism.

We briefly summarize the results of Stewart and Levin. The washout steady state $(S^0, 0, 0)$ is locally asymptotically stable if $f_u(S^0) < D$ and unstable if the reverse inequality holds. A unique plasmid-free steady state, $(\lambda, \bar{u}, 0)$ where $\bar{u} = \gamma(S^0 - \lambda)$ and $f_u(\lambda) = D$, exists only when the washout steady state is unstable, i. e. when $f_u(S^0) > D$. The plasmid-free steady state is stable if $\bar{\mu}\bar{u} < D + q - f_{u_+}(\lambda)$. They found that a unique coexistence steady state (S^*, u^*, u_+^*) will exist if:

$$\bar{\mu}\bar{u} > D + q - f_{u_+}(\lambda)$$

which can be rewritten as

$$\bar{\mu}\bar{u} > \chi D + q$$

where $\chi = 1 - \frac{f_{u_+}(\lambda)}{f_u(\lambda)}$. In the usual case that $\chi > 0$ when the plasmid-bearing population is at a growth disadvantage, the plasmid is maintained and the coexistence steady state exists if the density of plasmid-free organisms is sufficiently large relative to the cost of carrying the plasmid (χ) and the miss-segregation rate q. If

$$f_{u_+}(S) = f_u(S)(1-c) \qquad (11)$$

where c is the fractional energetic cost for plasmid carriage with $0 < c < 1$, the condition for coexistence becomes $\bar{\mu}\bar{u} > cD + q$.

The following result can be proved in a similar manner as those to follow.

Theorem 6.4.1 *Assume that (11) holds and $\gamma = \gamma_+$. Then the following hold for (10):*

(a) The washout state is globally stable whenever it is locally stable, which holds when $f_u(S^0) < D$.

(b) When $f_u(S^0) > D$, the plasmid-free steady state exists and it is asymptotically stable in the linear approximation if and only if $\bar{\mu}\bar{u} < cD + q$. In this case, it attracts all solutions with $u(0) + u_+(0) > 0$.

(c) When $\bar{\mu}\bar{u} > cD + q$ the unique coexistence steady state exists and attracts all solutions with $u_+(0) > 0$.

6.4.2 Stephanopoulus–Lapidus competition model

Stephanopoulus–Lapidus (1988) proposed a chemostat model of competition between plasmid-free and plasmid-bearing organisms which takes the form (4). It is based on earlier work of Ryder–DiBiasio (1984) who modeled segregation in a much different way than Stewart and Levin. They proposed that a fraction q of the daughter cells of the plasmid-bearing population produced in the time interval $[t, t+dt]$, given by $f_{u_+}(S)u_+ dt$, acquire no plasmid during cell division, and therefore contribute to the plasmid-free population, while the fraction $1 - q$ acquire one or more plasmid and thus contribute to the plasmid-bearing population. More precisely, of the daughter cells $f_{u_+}(S)u_+ dt$, $qf_{u_+}(S)u_+ dt$ are plasmid-free cells while $(1-q)f_{u_+}(S)u_+ dt$ are plasmid-bearing cells. This treatment of segregation seems to us more faithful to the biology since miss-segregation is associated with cell division. Cells don't lose plasmid, they just may not get one from the mother cell.

See also Hsu and Waltman (1997, 2004) for a similar approach in a different application.

The model of Stephanopoulus–Lapidus is given by

$$\begin{aligned} S' &= D(S^0 - S) - \gamma^{-1}[f_u(S)u + f_{u_+}(S)u_+] \\ u' &= (f_u(S) - D)u + qf_{u_+}(S)u_+ \\ u'_+ &= [f_{u_+}(S)(1-q) - D]u_+ \end{aligned} \qquad (12)$$

It has little to do with gene transfer so we include it here only because we adopt their approach to the modeling of segregation. The model is important in biotechnology where u_+ has been genetically engineered to produce some useful protein but miss-segregation implies that it must compete with the "wild-type" organism u. See the review article of Hsu and Waltman (2004) for more on models of this sort.

6.4.3 A model of gene transfer without wall growth

In this section we consider a model that is similar to the Stewart Levin model except that we employ the modeling of segregation introduced in (Ryder and DiBiasio 1984). Using (11), we are lead to consider the system

$$\begin{aligned} S' &= D(S^0 - S) - \gamma^{-1}f_u(S)[u + (1-c)u_+] \\ u' &= (f_u(S) - D)u + qf_u(S)(1-c)u_+ - \bar{\mu}uu_+ \\ u'_+ &= [f_u(S)(1-c)(1-q) - D]u_+ + \bar{\mu}uu_+ \end{aligned} \qquad (13)$$

where u and u_+ are the biomass concentrations of plasmid-free and plasmid-bearing organisms. As all the terms in (13) carry over from previous sections, no further motivation is needed. Observe that the plasmid bearing organism is assumed to have no advantage over the plasmid-free organism.

Adding all three equations in the above model and using the new variable $\Sigma = \gamma(S^0 - S) - u - u_+$ in place of S gives

$$\begin{aligned} \Sigma' &= -D\Sigma \\ u' &= (f_u(S) - D)u + qf_u(S)(1-c)u_+ - \bar{\mu}uu_+ \\ u'_+ &= [f_u(S)(1-c)(1-q) - D]u_+ + \bar{\mu}uu_+ \\ S &= S^0 - \gamma^{-1}(\Sigma + u + u_+) \end{aligned} \qquad (14)$$

It follows at once that $\Sigma(t) \to 0$, so the limiting system given by:

$$\begin{aligned} u' &= (f_u(S) - D)u + qf_u(S)(1-c)u_+ - \bar{\mu}uu_+ \\ u'_+ &= [f_u(S)(1-c)(1-q) - D]u_+ + \bar{\mu}uu_+ \\ S &= S^0 - \gamma^{-1}(u + u_+) \end{aligned} \qquad (15)$$

is the key to understanding the global dynamics of (13). We observe that in both (14) and (15), there are additional restrictions on the initial data aside from nonnegativity.

It is a routine exercise to show that solutions remain nonnegative. The ultimate boundedness of solutions of (13) is obvious from the fact that $\Sigma \to 0$.

Exactly as for the Stewart and Levin model, the steady states of (13) consist of a washout steady state $(S^0, 0, 0)$, a plasmid-free steady state $(\lambda, \bar{u}, 0)$ where $\bar{u} = \gamma(S^0 - \lambda)$ and a coexistence steady state denoted by (S^*, u^*, u_+^*).

We summarize our main results for (13).

Theorem 6.4.2 *The following hold:*

(a) the washout steady state is globally asymptotically stable whenever it is locally asymptotically stable and this occurs if and only if $f_u(S^0) < D$.

(b) When $f_u(S^0) > D$, the plasmid-free steady state exists and it is asymptotically stable in the linear approximation if and only if $f_u(\lambda)(1-c)(1-q) + \bar{\mu}\bar{u} < D$. It attracts all solutions with $u(0) + u_+(0) > 0$.

(c) When $f_u(\lambda)(1-c)(1-q) + \bar{\mu}\bar{u} > D$ then a unique coexistence equilibrium exists and attracts all solutions with $u_+(0) > 0$.

The proof of this theorem follows from Poincare–Bendixson Theorem and the following lemmas.

The stability of washout steady state of (13) can be determined by the eigenvalues of the variational matrix at $(S^0, 0, 0)$

$$J_1 := \begin{pmatrix} -D & -\gamma^{-1} f_u(S^0) & -\gamma^{-1} f_u(S^0)(1-c) \\ 0 & f_u(S^0) - D & q f_u(S^0)(1-c) \\ 0 & 0 & f_u(S^0)(1-c)(1-q) - D \end{pmatrix}$$

This leads immediately to the following result.

Lemma 6.4.3 *The washout steady state $(S^0, 0, 0)$ of (13) is locally asymptotically stable if and only if $f_u(S^0) < D$.*

Lemma 6.4.4 *If $f_u(S^0) < D$ then $u, u_+ \to 0$ as $t \to \infty$.*

Proof: Since $f_u(S^0) < D$, we can choose $\epsilon > 0$ small enough so that $f_u(S^0) + \epsilon < D$. From the first equation of (13)

$$S' \leq D(S^0 - S)$$

from which we conclude that $\limsup_{t \to \infty} S(t) \leq S^0$. Monotonicity of f_u implies that, for large enough t, $f_u(S(t)) \leq f_u(S^0) + \epsilon/2$, where ϵ is chosen above. Adding the last two equations of (13) and taking $v = u + u_+$ we have for large t,

$$\begin{aligned} v' &= (f_u(S) - D)u + f_u(S)(1-c)u_+ - Du_+ \\ &\leq 2(f_u(S) - D)v \\ &\leq 2(f_u(S^0) + \epsilon/2 - D)v \\ &\leq -\epsilon v. \end{aligned}$$

Since $u \geq 0$ and $u_+ \geq 0$, the result follows immediately. □

The criterion for stability of the plasmid-free steady state $(\lambda, \bar{u}, 0)$ of (13) is related to the variational matrix of (14) at the corresponding steady state $(0, \bar{u}, 0)$:

$$J_2 := \begin{pmatrix} -D & 0 & 0 \\ z & z & z + qD(1-c) - \bar{\mu}\bar{u} \\ 0 & 0 & D(1-c)(1-q) - D + \bar{\mu}\bar{u} \end{pmatrix}$$

where $z := -\gamma^{-1} f'_u(\lambda) \bar{u}$ and we have used that $f_u(\lambda) = D$.

Lemma 6.4.5 *The plasmid-free steady state $(\lambda, \bar{u}, 0)$ of (13) is stable if and only if $\bar{\mu}\bar{u} < D[1 - (1-c)(1-q)]$.*

Proof: The variational matrix J_2 of (14) has eigenvalues $-D$ and the eigenvalues of the lower right two-by-two sub-matrix. The eigenvalues of J_2 have negative real parts when the above stated condition is satisfied. □

The plasmid-free steady state $(\lambda, \bar{u}, 0)$ is unstable if

$$\frac{\bar{\mu}\bar{u}}{D} + \frac{f_u(\lambda)(1-c)(1-q)}{D} > 1. \tag{16}$$

The first term on the left gives the number of infections produced by a single plasmid-bearing cell in the environment determined by the plasmid-free steady state before being washed out. The second term gives the number of plasmid-bearing daughter cells of a single plasmid-bearing cell before washing out. Of course, the factor $f_u(\lambda)/D = 1$ but we leave it in for interpretations sake. Thus, the sum gives the number of horizontal and vertical transmissions of the plasmid before washout. That number must exceed one for plasmid persistence.

Lemma 6.4.6 *There exists a unique coexistence steady state (S^*, u^*, u^*_+) where $S^0 > S^* > \lambda$ when the plasmid-free steady state is unstable. There can be no coexistence steady state when it is stable.*

Proof: Adding the three steady state equations for (13) gives

$$\gamma(S - S^0) = u + u_+ \tag{17}$$

Solving the second and third equation for u and u_+, we get

$$u = \frac{D - f_u(S)(1-c)(1-q)}{\bar{\mu}}$$

$$u_+ = \frac{[D - f_u(S)]}{f_u(S)(1-c) - D} \frac{D - f_u(S)(1-c)(1-q)}{\bar{\mu}}$$

Substituting these into (17) and a little algebra leads to a single equation for S:

$$\bar{\mu}\gamma(S^0 - S) = cf_u(S)[1 + \frac{qf_u(S)(1-c)}{D - f_u(S)(1-c)}]$$

Positivity of u, u_+ implies that we must have $f_u(S) > D$ and $f_u(S)(1-c) < D$; clearly, $0 < S < S^0$. Let $F(S)$ denote the left hand side and $G(S)$ denote the right hand side of the equality. F is obviously decreasing. The term in square brackets in G is a monotonically increasing function of S and is positive when $D - f_u(S)(1-c) > 0$. Thus $G(S)$ is a monotonically increasing function of S so long as $D - f_u(S)(1-c) > 0$ and it satisfies $G(0) = 0$, $G(\lambda) = D[1 - (1-c)(1-q)]$ since $f_u(\lambda) = D$. Thus, there is at most one value of S where $F(S) = G(S)$ on the interval where $D - f_u(S)(1-c) > 0$. Note that $F(\lambda) = \bar{\mu}\bar{u}$. If the plasmid-free state is hyperbolically stable then $F(\lambda) < G(\lambda)$ so the intermediate value theorem gives the unique value $S^* \in (0, \lambda)$ where $F = G$. But $f_u(S^*) < f_u(\lambda) = D$ implying that u and u_+ are not both positive. There exists no coexistence steady state when the plasmid-free state is hyperbolically stable. Similarly, when $\bar{\mu}\bar{u} = D[1 - (1-c)(1-q)]$ we get the same contradiction. When $\bar{\mu}\bar{u} > D[1 - (1-c)(1-q)]$, that is, when the plasmid-free steady state is unstable, then $F(\lambda) > G(\lambda)$ so $S^* > \lambda$ if it exists. There are two cases depending on whether $D - f_u(S^0)(1-c) > 0$ or $D - f_u(S^0)(1-c) \leq 0$. In the first case, $F(S^0) = 0 < G(S^0)$ so $S^* \in (\lambda, S^0)$ exists by the intermediate value theorem. In the second case, G has a vertical asymptote at some $\tilde{S} \leq S^0$ and in this case too the intermediate value theorem implies the existence of $S^* \in (\lambda, \tilde{S})$. Since $D - f_u(S^*) < D - f_u(\lambda) = 0$, the values of u and u_+ above are positive. □

The local stability of the coexistence steady state (S^*, u^*, u_+^*) of (13) can be determined from the eigenvalues of the variational matrix of (14) at its corresponding steady state $(0, u^*, u_+^*)$

$$J_3 := \begin{pmatrix} -D & 0 & 0 \\ \cdot & j_{22} & j_{23} \\ \cdot & j_{32} & j_{33} \end{pmatrix}$$

where

$$j_{22} = f_u(S^*) - D - \gamma^{-1}u^* f_u'(S^*) - q\gamma^{-1}u_+^* f_u'(S^*)(1-c) - \bar{\mu}u_+^*$$
$$j_{23} = -\gamma^{-1}u^* f_u'(S^*) + qf_u(S^*)(1-c) - \gamma^{-1}qu_+^* f_u'(S^*)(1-c) - \bar{\mu}u^*$$
$$j_{32} = -\gamma^{-1}u_+^* f_u'(S^*)(1-c)(1-q) - \bar{\mu}u_+^*$$
$$j_{33} = -\gamma^{-1}u_+^* f_u'(S^*)(1-c)(1-q)$$

Note that the entries denoted "·" play no role in the stability of coexistence steady state. In the remainder of the proof, we use the notation

$$d = 1 - c, \quad p = 1 - q$$

in order to shorten lengthy formulae. The coexistence steady state of (13) is stable if the eigenvalues of the matrix $E = (j_{jk})_{j,k \in \{2,3\}}$ have negative real parts i. e. trace$(E) < 0$ and det$(E) > 0$. Since $\bar{\mu}u_+^* > f_u(S^*) - D$, $j_{22} < 0$ and since $j_{33} < 0$, trace$(E) < 0$. In order to show that det$(E) > 0$ we simplify j_{23} as follows:

$$\begin{aligned}j_{23} &= -\gamma^{-1}u^* f_u'(S^*) + qf_u(S^*)d - q\gamma^{-1}u_+^* f_u'(S^*)d - \bar{\mu}u^* \\ &= -\gamma^{-1}u^* f_u'(S^*) + qf_u(S^*)d - q\gamma^{-1}u_+^* f_u'(S^*)d \\ &\quad - D + f_u(S^*)dp \\ &= -\gamma^{-1}u^* f_u'(S^*) - q\gamma^{-1}u_+^* f_u'(S^*)d - D + f_u(S^*)d \\ &< 0.\end{aligned}$$

because the sum of the last two terms is negative. If $\bar{\mu} \geq \gamma^{-1}f_u'(S^*)dp$ then $j_{32} \geq 0$ so $det(E) > 0$ while if $\bar{\mu} < \gamma^{-1}f_u'(S^*)dp$ then

$$\begin{aligned}\det(E) &= -\gamma^{-1}u_+^* f_u'(S^*)dp f_u(S^*) + \gamma^{-1}u_+^* f_u'(S^*)dpD \\ &\quad + \gamma^{-1}u_+^* f_u'(S^*)dpq f_u(S^*)d + \bar{\mu}\gamma^{-1}(u_+^*)^2 f_u'(S^*)dp \\ &\quad - \bar{\mu}u^* \gamma^{-1}u_+^* f_u'(S^*)dp + \bar{\mu}u_+^* \gamma^{-1}u^* f_u'(S^*) \\ &\quad - \bar{\mu}u_+^* q f_u(S^*)d + \bar{\mu}u_+^* \gamma^{-1}u_+^* q f_u'(S^*)d + \bar{\mu}u^* \bar{\mu}u_+^* \\ &= \gamma^{-1}f_u'(S^*)dpu_+^*[\bar{\mu}u_+^* + D - f_u(S^*)] \\ &\quad + qf_u(S^*)du_+^*[\gamma^{-1}f_u'(S^*)dp - \bar{\mu}] \\ &\quad + \gamma^{-1}f_u'(S^*)\bar{\mu}u^* u_+^*[1 - dp] + \gamma^{-1}u_+^* q f_u'(S^*)d + \bar{\mu}u^* \bar{\mu}u_+^* \\ &> 0.\end{aligned}$$

The above inequality is true because all three terms inside the square brackets are positive. Thus we have, trace$(E) < 0$ and det$(E) > 0$ and so (S^*, u^*, u_+^*) is locally asymptotically stable.

Lemma 6.4.7 *System (15) has no periodic solutions.*

Proof of Lemma 6.4.7: We apply the Dulac criterion with the auxiliary function

$$g(u, u_+) = \frac{1}{uu_+}$$

to the system (15) and find that

$$\frac{\partial}{\partial u}[g(u, u_+)u'] + \frac{\partial}{\partial u_+}[g(u, u_+)u_+']$$
$$= -\frac{\gamma^{-1}f_u'(S)}{u_+} - q(1-c)\frac{\gamma^{-1}uf_u'(S) + f_u(S)}{u^2} - \frac{\gamma^{-1}f_u'(S)(1-c)(1-q)}{u} < 0.$$

Hence, the Dulac criterion implies that the above system does not have any periodic solution. □

Proof of Theorem 6.4.2: Part (a) follows from Lemma 6.4.4. For parts (b) and (c), we first consider the planer system (15). If $f_u(S^0) > D$ and $f_u(\lambda)(1-c)(1-q) + \bar{\mu}\bar{u} < D$, there are two steady states: the washout state $(0,0)$ is unstable and the plasmid-free state $(\bar{u}, 0)$ is asymptotically stable and it is attracts all orbits with $u(0) > 0$ by the Poincare–Bendixson theorem and Lemma 6.4.7. If If $f_u(S^0) > D$ and $f_u(\lambda)(1-c)(1-q) + \bar{\mu}\bar{u} > D$, both the washout and the plasmid-free states are unstable and the coexistence state (u^*, u_+^*) is stable. Again, by the Poincare–Bendixson theorem and Lemma 6.4.7, the coexistence state attracts all orbits with initial condition $u_+(0) > 0$.

Now we consider the system (14). For case (b) and (c), all steady states are hyperbolic under our hypotheses so hypotheses $(H1) - (H4)$ of theorem (F.1) of [17] are satisfied. There are no cycles of equilibria, so $(H5)$ is also satisfied. Theorem (F.1) of (Smith and Waltman 1995) implies that those trajectory identified in cases (b) and (c) tend to the locally asymptotically stable steady state. □

Figure 6.2 depicts the invasion of the plasmid-free state by a tiny inoculum of plasmid-bearing organisms. We use (6) for growth and uptake. The output has been scaled by S/a, $u/(a\gamma)$ and $u_+/(a\gamma)$. Parameter values are chosen as in Freter (1983), as used in Jones et al. (2002). In particular, $\gamma = 0.5$,

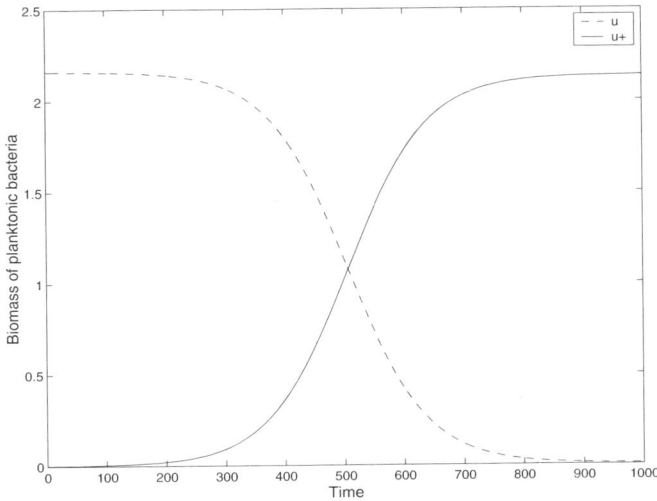

Fig. 6.2. Time series of the invasion of the plasmid-free steady state by an inoculum of plasmid-bearing organisms with $\bar{\mu} = .0018 \times 10^7$

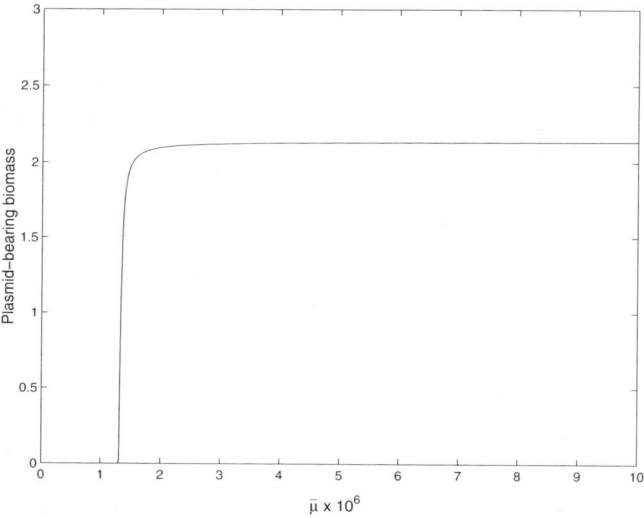

Fig. 6.3. Bifurcation diagram depicting the coexistence steady state value of u_+ versus $\bar{\mu}$

$a = 9 \times 10^{-7}$ g/ml, $m = 1.66\,\text{hr}^{-1}$, $S^0 = 2.09 \times 10^{-6}$ g/ml, $D = 0.23\,\text{hr}^{-1}$, $V = 1\,\text{cm}^3$, $A = 6\,\text{cm}^2$. Simonsen (1991) suggests $c = 0.01$ and $q = 0.0001$. He also points out that the value of $\bar{\mu}$ ($a\gamma\bar{\mu}$ with current scaling) is highly uncertain. We take $\bar{\mu} = .0018 \times 10^7$, a factor of 10^7 larger than biologically reasonable, in order to satisfy condition (c) of Theorem 6.4.2.

Initial data are chosen to be near the plasmid-free steady state $(\lambda, \bar{u}) = (0.16084, 2.1614)$ with S, u exactly at steady state and $u_+ = 0.001$.

Figure 6.3 plots the coexistence value of u_+ versus the conjugational transfer parameter $\bar{\mu}$. A very large value of $\bar{\mu}$ is required for the persistence of the plasmid-bearing organism, reflecting our assumption that the plasmid confers no advantage on its host.

6.5 A model of gene transfer in biofilms

In this section we obtain our main results concerning (1), restated below for the convenience of the reader.

$$\begin{aligned}
S' &= D(S^0 - S) - \gamma^{-1} f_u(S)\left[u + (1-c)u_+\right] - \gamma^{-1}\delta f_w(S)\left[w + (1-c)w_+\right] \\
u' &= (f_u(S) - D)u + q f_u(S)(1-c)u_+ - \alpha u + \beta\delta w - \bar{\mu}u u_+ \\
w' &= f_w(S)w + q f_w(S)(1-c)w_+ + \alpha\delta^{-1} u - \beta w - \mu w w_+ \\
u'_+ &= [f_u(S)(1-c)(1-q) - D]u_+ - \alpha_+ u_+ + \beta_+ \delta w_+ + \bar{\mu}u u_+ \\
w'_+ &= f_w(S)(1-c)(1-q)w_+ + \alpha_+ \delta^{-1} u_+ - \beta_+ w_+ + \mu w w_+
\end{aligned} \qquad (18)$$

Key features of the model are summarized as follows:

1. growth and uptake rates of the plasmid-bearing organism are a factor $1-c$ lower than those for the plasmid-free organism reflecting the cost of bearing plasmid.
2. adhering and sloughing rates for plasmid-bearing (α_+, β_+) and plasmid-free organism (α, β) may differ.
3. fraction q of daughter cells of plasmid-bearing cells do not receive plasmid.
4. plasmid-bearing organisms transmit plasmid via conjugation to plasmid-free organisms in both fluid and wall environments, though perhaps at different rates $(\bar{\mu} \neq \mu)$.
5. all yield coefficients have been taken to be the same (γ).

The model differs from the one in (Imran et al. preprint) where the plasmid-bearing organism's growth rate, but not its uptake rate, was assumed to be reduced by a factor $1-c$. See the discussion section for an elaboration of this difference.

Our main focus is on conditions under which the plasmid-bearing organism, whose densities are given by u_+, w_+ can survive. The set $u_+ = w_+ = 0$, where they are absent, is invariant and the equations describing the dynamics on this subset are

$$S' = D(S^0 - S) - \gamma^{-1}[f_u(S)u + f_w(S)\delta w]$$
$$u' = (f_u(S) - D)u - \alpha u + \beta \delta w$$
$$w' = f_w(S)w + \alpha \delta^{-1}u - \beta w \tag{19}$$

We refer to it as the plasmid-free system, noting that it is identical to (5). Our main assumptions concern (19) and are collected in the following:

(H) The washout state $(S^0, 0, 0)$ is unstable for (19) (i.e., $s(B(S^0)) > 0$, see (8)) and the survival state $(\bar{S}, \bar{u}, \bar{w})$ attracts all solutions of (19) satisfying $u(0) + \delta w(0) > 0$.

As noted in Theorem 6.3.1 (d), (H) holds when $f_u = f_w$ and $R_0 > 1$. We ignore the case that the washout state for (19) is stable because then it is a global attractor for (19) by Theorem 6.3.1 (a) and, we conjecture, also for (18) although we do not yet have a proof of this.

The structure of (18) implies a restriction on the types of steady states. Obviously, we have the washout state $(S^0, 0, 0, 0, 0)$ and, we will show that the plasmid-free state $(\bar{S}, \bar{u}, \bar{w}, 0, 0)$ exists when the washout state is unstable. However, there is no comparable "plasmid-bearing" state because the segregational loss of plasmid guarantees that where there are plasmid-bearing cells, there will be plasmid-free cells. Especially important are possible coexistence states $(S^*, u^*, w^*, u_+^*, w_+^*)$ which imply plasmid persistence.

The key question is whether or not the plasmid-bearing population can invade the plasmid-free steady state $(\bar{S}, \bar{u}, \bar{w}, 0, 0)$ leading to the persistence of the plasmid. The answer comes from the linearization of (18) about the plasmid-free state, the jacobian matrix of which, takes the form

$$J := \begin{pmatrix} J_{3\times 3} & X_{3\times 2} \\ 0_{2\times 3} & C_{2\times 2} \end{pmatrix}$$

where $0_{2\times 3}$ is the zero matrix and $J_{3\times 3}$ is a stable matrix because the plasmid-free state is asymptotically stable for (19) by Theorem 6.3.1. Thus, the stability of the plasmid-free state is determined by the eigenvalues of the sub-matrix C, given by:

$$\begin{pmatrix} f_u(\bar{S})dp - D - \alpha_+ + \bar{\mu}\bar{u} & \beta_+ \\ \alpha_+ & f_w(\bar{S})dp - \beta_+ + \mu\bar{w} \end{pmatrix} \qquad (20)$$

where $d = 1 - c$ and $p = 1 - q$. If the stability modulus, $s(C)$, of C (the largest eigenvalue) is negative then the plasmid-free state is locally attracting; if $s(C) > 0$ then the plasmid-free state is unstable. In this case, the plasmid is maintained.

Theorem 6.5.1 *Assume hypothesis (H) holds. If $s(C) > 0$ then the plasmid-bearing population persists. More precisely, there exists $\epsilon > 0$, independent of initial data, such that for all solutions of (18) satisfying $u_+(0) + \delta w_+(0) > 0$, we have*

$$u_+(t) + \delta w_+(t) > \epsilon \qquad (21)$$

for all sufficiently large t. In addition, there is at least one coexistence steady state:

$$(S^*, u^*, w^*, u_+^*, w_+^*)$$

with positive components.

Figure 6.4 depicts the invasion of the plasmid-free state by a tiny inoculum of plasmid-bearing organisms. The output has been scaled by S/a, $u/(a\gamma)$, $\delta w/a\gamma$ and similarly for u_+, w_+. Parameter values are the same used in previous section except for μ, which is taken as in (Imran et al. preprint). Initial data are chosen to be near the plasmid-free steady state $(\bar{S}, \bar{u}, \bar{w}, 0, 0) = (.11, 2.21, .93, 0, 0)$ with S, u and w exactly at steady state and $u_+ = 0.001$, $w_+ = 0$. Observe that the simulation tracks the plasmid-free steady state for the first 60 hours then makes a transition to a coexistence state dominated by wall-adherent, plasmid-bearing cells.

Local existence and positivity of solutions of (18) are standard (see Smith and Waltman 1995). A key to proving Theorem 6.5.1 is establishing a uniform ultimate upper bound on solutions.

Lemma 6.5.2 *All nonnegative solutions of (18) are ultimately uniformly bounded in forward time, and thus they exist for all positive time. In fact,*

$$\limsup_{t\to\infty}(S + \frac{u}{\gamma} + \frac{\delta w}{\gamma} + \frac{u_+}{\gamma} + \frac{\delta w_+}{\gamma}) \leq S^0/b \qquad (22)$$

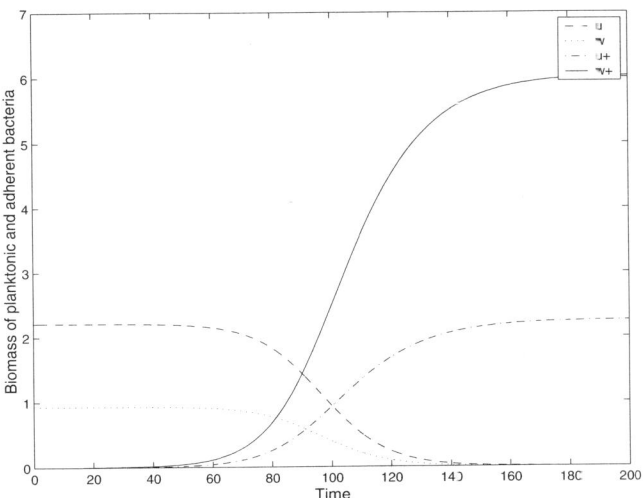

Fig. 6.4. Time series of the invasion of the plasmid-free steady state by inoculum of plasmid-bearing organisms that are better adherers. $\alpha = \alpha_+ = \beta = .1, \beta = .4, \mu = 1$ and $\bar{\mu} = .0018$

where $\bar{\beta} = min\{\beta, \beta_+\}$, $d = max\{D + \alpha + \bar{\beta} + f_w(S^0), f_u(S^0)c + D + \alpha_+ + \bar{\beta} + f_w(S^0)\}$, $e = f_u(S^0)$, and $b = \frac{-d + \sqrt{d^2 + 4\bar{\beta}e}}{2e}$.

Proof: From the inequality

$$S' \leq D(S^0 - S)$$

we conclude that $\limsup_{t\to\infty} S \leq S^0$. Monotonicity of f_u and f_w imply that, for given $\epsilon > 0$ we have $f_u(S(t)) \leq f_u(S^0) + \epsilon$ and $f_w(S(t)) \leq f_w(S^0) + \epsilon$, for $t \geq T$

For given a solution we define

$$M(t) = \frac{u + u_+}{u + u_+ + \delta w + \delta w_+}.$$

Then

$$M' = [\frac{(u + u_+)'}{(u + u_+ + \delta w + \delta w_+)}] - [\frac{(u + u_+ + \delta w + \delta w_+)'(u + u_+)}{(u + u_+ + \delta w + \delta w_+)^2}] =: l - n.$$

The first square bracket, l, is

$$l = \frac{(f_u u - Du - \alpha u + \beta \delta w + f_u(1-c)u_+ - Du_+ - \alpha_+ u_+ + \beta_+ \delta w_+)}{(u + u_+ + \delta w + \delta w_+)},$$

where $f_u = f_u(S(t))$ and $f_w = f_w(S(t))$. If $a = a(t) := \min\{(f_u - D - \alpha), (f_u(1-c) - D - \alpha_+)\}$, then

$$l \geq aM + \bar{\beta}(1-M).$$

The second square bracket, n, in M' is

$$n = -\frac{(f_u u - Du + f_w \delta w + f_u(1-c)u_+ - Du_+ + (f_w(1-c)\delta w_+)(u+u_+)}{(u+u_+ + \delta w + \delta w_+)^2}$$

$$n \geq \frac{(-f_u u - f_u u_+ - f_w \delta w - f_w \delta w_+)}{(u + u_+ + \delta w + \delta w_+)} M + DM^2$$

$$n \geq -f_w M - f_u M^2 + f_w M^2 + DM^2$$

So

$$M' \geq \bar{\beta} + (a - \bar{\beta} - f_w)M + M^2(D + f_w - f_u).$$

Using the result of the first paragraph of the proof, and considering both cases one by one for a, given $\epsilon > 0$, there is $T > 0$ such that

$$a - \bar{\beta} - f_w \geq -d - \epsilon + f_u$$
$$\geq -d - \epsilon$$

for all $t \geq T$. So $M' \geq \bar{\beta} - M(d+\epsilon) - M^2 e/2$. The right hand side of this inequality is a parabola opening down wards. Inside the positive region there is only one stable rest point. Consequently,

$$\frac{-(d+\epsilon) + \sqrt{(d+\epsilon)^2 + 2\bar{\beta}e}}{e} \leq \liminf_{t\to\infty} M$$

and since $\epsilon > 0$ is arbitrary,

$$\frac{-d + \sqrt{d^2 + 2\bar{\beta}e}}{e} \leq \liminf_{t\to\infty} M$$

Let $z = S + \frac{u}{\gamma} + \frac{\delta w}{\gamma} + \frac{u_+}{\gamma} + \frac{\delta w_+}{\gamma}$. Adding the five equations of (1) we find that,

$$z' = D(S^0 - S - \frac{u}{\gamma} - \frac{u_+}{\gamma})$$

For ϵ satisfying $b = \frac{-d+\sqrt{d^2+2\bar{\beta}e}}{e} > \epsilon > 0$, there exists $T > 0$ such that for $t \geq T$

$$u + u_+ \geq [b - \epsilon](u + u_+ + \delta w + \delta w_+).$$

Therefore, for $t \geq T$

$$z' \leq D(S^0 - S) - D\gamma^{-1}(b-\epsilon)(u + u_+ + \delta w + \delta w_+)$$
$$\leq D[S^0 - (b-\epsilon)z]$$

implying that $\limsup_{t \to \infty} z \leq S^0/b$. □

Note that it is critical for the proof that $\beta, \beta_+ > 0$. If, for example, $\beta = 0$, the wall population may grow unboundedly.

Proof of Theorem 6.5.1: We follow a similar argument used in Theorem 5.3 of (Stemmons and Smith 2000), applying Theorem 4.6 in (Thieme 1993). Lemma 22 establishes that (18) has a compact attractor so that the dissipativity requirement of Theorem 4.6 holds.

Using the notation of that result, we set $X = R_+^5$, $X_2 = \{(S, u, w, u_+, w_+) \in X : u_+ = 0 \text{ or } w_+ = 0\}$, and $X_1 = X \setminus X_2$. Observe that solutions of (18) starting in X_2 immediately enter X_1, where $u_+, w_+ > 0$, unless $u_+(0) = w_+(0) = 0$. We want to show that solutions which start in X_1 are eventually bounded away from X_2. Using the notation $x(t) = (S(t), u(t), w(t), u_+(t), w_+(t))$ for a solution of (18), define

$$Y_2 = \{x(0) \in X_2 : x(t) \in X_2, t \geq 0\} = \{x(0) \in X : u_+(0) = w_+(0) = 0\}$$

and Ω_2, the union of omega limit sets of solutions starting in X_2, is, by our hypotheses, the set $\{E_0, E_1\}$ where $E_0 := (S^0, 0, 0, 0, 0)$ and $E_1 := (\bar{S}, \bar{u}, \bar{w}, 0, 0)$. We will show that if $M_0 = \{E_0\}$ and $M_1 = \{E_1\}$, then $\{M_0, M_1\}$ is an isolated acyclic covering of Ω_2 in Y_2 and each M_i is a weak repeller. All solutions starting in Y_2 but not on the S-axis converge to E_1 while those on the axis converge to E_0. E_1, being locally asymptotically stable relative to Y_2, cannot belong to the alpha limit set of any full orbit in X_2 different from E_1 itself. Similar arguments apply to E_0; the only solutions converging to it lie on the S-axis and these are either unbounded or leave X in backward time. Thus $\{M_0, M_1\}$ is an acyclic covering of Ω_2. If M_1 were not a weak repeller for X_1, there would exist an $x(0) \in X_1$ such that $x(t) \to E_1$ as $t \to \infty$. Let $V(t) = (u_+(t), \delta w_+(t))^t$ and define the matrix $P(f(S), u, w)$ ($f = (f_u, f_w)$) by

$$\begin{pmatrix} f_u(S)(1-c)(1-q) - D - \alpha_+ + \bar{\mu}u & \beta_+ \\ \alpha_+ & f_w(S)(1-c)(1-q) - \beta_+ + \mu w \end{pmatrix} \tag{23}$$

Then $A = P(f(\bar{S}), \bar{u}, \bar{w})$ and we may write the equation satisfied by $V(t)$ as

$$\dot{V} = P(f(\bar{S}), \bar{u}, \bar{w})V + [P(f(S), u, w) - P(f(\bar{S}), \bar{u}, \bar{w})]V$$

If $P(f(\bar{S}), \bar{u}, \bar{w})^t W = qW$ where $q = s(P(f(\bar{S}), \bar{u}, \bar{w})) = s(C) > 0$ and $W = (m, n)^t$ with $m, n > 0$ is the Perron–Frobenius eigenvector, then on taking the scalar product of both sides of the differential equation by W and using that $S(t) \to \bar{S}$ and $w(t) \to \bar{w}$, we have

$$\frac{d}{dt}(mu_+ + n\delta w_+) \geq q/2(mu_+ + n\delta w_+)$$

for all large t. But this leads to the contradiction to $x(t) \to E_1$, namely that $mu_+(t) + n\delta w_+(t) \to \infty$ as $t \to \infty$. Thus M_1 is a weak repeller. The argument above together with the fact that E_1 is locally asymptotically stable relative to the subspace $(u_+, w_+) = (0, 0)$ implies that it is an isolated compact invariant set in X. Similar arguments show that M_0 is a weak repeller and an isolated compact invariant set in X. Therefore, Theorem 4.6 in (Thieme 1993) implies our result: there exists $\epsilon > 0$ such that $\liminf_{t \to \infty} d(x(t), X_2) > \epsilon$ for all $x(0) \in X_1$, where $d(x, X_2)$ is the distance from x to X_2.

The existence of at least one coexistence steady state follows from Theorem 1.3.7 of (Zhao 2003). □

6.6 Discussion

Building on previous work, we have constructed a model of gene transfer between micro-organisms in a heterogeneous environment consisting of a biofilm immersed in a fluid medium. The chemostat setting of our model may not be appropriate in many natural environments so we point out here how the model can be modified for different settings (but see also (Imran et al. preprint) for a spatially explicit setting). Equations (1) reflect the chemostat mainly due to the fact that the same term D serves simultaneously as the input rate of supply of fresh substrate, the outflow rate of unused substrate and the removal rates of planktonic cells, both u and u_+. If one replaces the removal rates of planktonic cells by a parameter D', possibly distinct from D, then (1) is extended in a way that may better capture natural environments. Even in the chemostat setting, one may view D' as $D + d$ where d is a death rate of bacteria. Our analysis continues to hold although one must modify slightly Lemma 5.2 and note that the quoted results of Pilyugin and Waltman (1999) have not been established in this setting.

Our model (1) represents a slight modification of the one originally proposed in (Imran et al. preprint). In (Imran et al. preprint), it was assumed that plasmid bearing organisms have the same nutrient uptake rate as plasmid-free organisms but their growth rates are reduced by a factor $1 - c$; bearing plasmid negatively affects growth but not uptake. This implies that the effective yield of plasmid-bearing organisms is reduced by this same factor. In the present work, we follow previous workers by assuming that both uptake and growth of plasmid-bearing organisms are reduced by the factor $1 - c$ so the yield remains the same. In other words, it is assumed that bearing plasmid reduces both uptake and growth rates. This assumption has the effect of greatly improving the mathematical tractability of the model of gene transfer without wall growth (because a conservation relation holds) but does not significantly affect the analysis of (1). It is likely that uptake and growth rate are affected differently and that the magnitude of each effect depends both on the particular microorganism and on the particular plasmid.

In order to better understand our model, we have reviewed previous work where the key modeling ideas were first developed. These include the work of Levin and Stewart from which most of the mathematical modeling of plasmid transfer can be traced, work of Ryder and DiBiasio (1984) and Stephanopolis and Lapidus (1988) where a more realistic modeling of miss-segregation was proposed, and the work of Pilyugin and Waltman (1999) whose simple biofilm model forms the basis of our model. Our focus in the present chapter, as in these earlier works, is on understanding the conditions that allow the plasmid to persist in a bacterial population despite conferring a growth disadvantage to its bearer and despite the occasional leakage in its vertical transmission from mother to daughter cells. In each of the models considered here, the key to understanding plasmid persistence lies in determining the basic reproduction number R_0, the number of plasmid-bearing progeny that a hypothetical single plasmid-bearing cell would leave if introduced into the plasmid-free steady state environment. These progeny consist of daughter cells born with plasmid and formerly plasmid-free organisms that have acquired the plasmid via conjugation. In the chemostat setting of these models, a cell eventually washes out so a key quantity involved in the calculation of R_0 is the mean residence time (MRT) in the chemostat. Plasmid persistence requires that $R_0 > 1$.

The Stewart and Levin model (10) is chemostat based so MRT $= 1/D$ but a plasmid-bearing cell reverts to a plasmid-free cell at rate q so the mean time our plasmid-bearing cell remains in the chemostat and remains plasmid-bearing is $1/(D+q)$. This leads to

$$R_0 = [f_{u_+}(\lambda) + \bar{\mu}\bar{u}]/(D+q)$$

The term $f_{u_+}(\lambda)/(D+q)$ gives the number of daughter cells born to the single plasmid-bearing cell before washout. The term $\bar{\mu}\bar{u}/(D+q)$ gives the number of plasmid-free cells infected by the plasmid-bearing cell before it washes out. Consequently, the condition $R_0 > 1$ for plasmid persistence just says that the number of plasmid-bearing progeny must exceed one.

Our model of gene transfer without wall growth (13) contains the more realistic modeling of miss-segregation developed by Ryder and DiBiasio (1984) and Stephanopolis and Lapidus (1988). This model does not allow a plasmid-bearing cell to revert to a plasmid-free cell-only one of its daughter cells can be plasmid free. In this case

$$R_0 = [f_u(\lambda)(1-c)(1-q) + \bar{\mu}\bar{u}]/D$$

The interpretation is similar to that above since $f_{u_+}(S) = f_u(S)(1-c)$ but one must discount the progeny $qf_u(\lambda)(1-c)/D$ of our single plasmid-bearing cell that do not carry the plasmid.

We would like to determine the basic reproductive number for our gene transfer model in a biofilm (Berman and Plemmons 1979). How are we to reinterpret the condition for plasmid persistence

$$s(C) > 0$$

in biological terms? As we will see, the Perron–Frobenius theory (see [2]) gives estimates of $s(C)$ which are biologically interpretable. Let

$$Q_+ := \begin{pmatrix} -D - \alpha_+ & \beta_+ \\ \alpha_+ & -\beta_+ \end{pmatrix}$$

In the appendix we show that the mean residence time of a plasmid-bearing cell in the chemostat is given by

$$\mathrm{MRT}_+ = -1/s(Q_+)$$

Define

$$k := \min\{f_u(\bar{S})dp + \bar{\mu}\bar{u}, f_w(\bar{S})(1-c)(1-q) + \mu\bar{w}\}$$
$$K := \max\{f_u(\bar{S})dp + \bar{\mu}\bar{u}, f_w(\bar{S})(1-c)(1-q) + \mu\bar{w}\}$$

Then

$$kI + Q_+ \leq C \leq KI + Q_+$$

which implies that

$$k + s(Q_+) = s(kI + Q_+) \leq s(C) \leq s(KI + Q_+) = K + s(Q_+).$$

Consequently, we may express $s(C)$ as follows

$$s(C) = F_+[f_u(\bar{S})dp + \bar{\mu}\bar{u}] + (1 - F_+)[f_w(\bar{S})dp + \mu\bar{w}] + s(Q_+)$$

for some F_+ with $0 \leq F_+ \leq 1$. The condition $0 < s(C)$ for plasmid persistence can then be equivalently expressed as $0 < -s(C)/s(Q_+)$, or as

$$1 < F_+[f_u(\bar{S})dp + \bar{\mu}\bar{u}] + (1 - F_+)[f_w(\bar{S})dp + \mu\bar{w}]) \cdot \mathrm{MRT}_+ \qquad (24)$$

Inequality (24) can be viewed as requiring that a single plasmid-bearing cell, introduced into the plasmid-free steady state, leave more than one plasmid-bearing progeny in order for persistence of the plasmid. Indeed, the term

$$f_u(\bar{S})dp \cdot \mathrm{MRT}_+$$

gives the number of daughter cells carrying plasmid born of the single plasmid-bearing cell before it washes out of the chemostat, assuming that it resides in the fluid during this time. Replacing the first factor by $f_w(\bar{S})dp$

gives the corresponding number of plasmid-bearing progeny assuming that the cell spends its time adhering to the wall. The term

$$\bar{\mu}\bar{u} \cdot \mathrm{MRT}_+$$

gives the number of infectious transfers of plasmid from our single plasmid-bearing cell to plasmid-free cells before washout, assuming that it resides in the fluid during this time. Replacing the first factor by $\mu\bar{w}$ gives the corresponding number of infectious tranfers of plasmid given that the cell resides on the wall during its time in the chemostat.

But we must take into account that our lone plasmid-bearing cell, introduced into the plasmid-free steady state, will spend a certain fraction F_+ of its residence time in the fluid and a complementary fraction $1 - F_+$ on the wall. In this way, the interpretation of (24) is clear-a plasmid-bearing cell must leave more than one plasmid-bearing progeny.

Diekmann and Heesterbeek (2000) (see Thm 6.13) show that

$$s(C) > 0 \Leftrightarrow \rho(-TQ_+^{-1}) > 1$$

where $\rho(A)$ denotes the spectral radius of A,

$$T = \mathrm{diag}[f_u(\bar{S})d\bar{p} + \bar{\mu}\bar{u}, f_w(\bar{S})d\bar{p} + \mu\bar{w}]$$

and

$$-Q_+^{-1} = \begin{pmatrix} 1/D & 1/D \\ \alpha_+/D\beta_+ & (\alpha_+ + D)/D\beta_+ \end{pmatrix}$$

See (Imran et al. preprint) for a discussion of the biological meaning for the entries of this matrix.

Thus they are lead to define the basic reproductive number

$$R_0 = \rho(-TQ_+^{-1}) .$$

We observe that

$$\rho(-Q_+^{-1}) = -1/s(Q_+) = \mathrm{MRT}_+$$

where the subscript "+" denotes that the plasmid-bearing organisms parameters α_+ and β_+ are used. Using the Perron–Frobenius theory, with $\rho(-TQ_+^{-1})$ instead of $s(C)$, as before leads to

$$R_0 = (F_+[f_u(\bar{S})d\bar{p} + \bar{\mu}\bar{u}] + (1-F_+)[f_w(\bar{S})d\bar{p} + \mu\bar{w}]) \cdot \mathrm{MRT}_+ \qquad (25)$$

Let's return to the biology and first ask whether the plasmid could survive if it confers no advantage in biofilm forming ability on its host. That is, assuming:

$$\alpha = \alpha_+ , \quad \beta = \beta_+ ,$$

can the plasmid survive as a parasite. This question was considered by Stewart and Levin (1977) for the simple chemostat-based model ignoring wall growth. The question boils down to whether or not its advantage in horizontal spread outweighs its growth and segregation disadvantages. In this case, both plasmid-bearing and plasmid-free cells have identical mean residence times in the chemostat, MRT = MRT$_+$. In order to interpret formula (24) in this case, consider that at the plasmid-free steady state each plasmid-free organism leaves exactly one daughter cell. Thus

$$F \cdot f_u(\bar{S}) \cdot \text{MRT} + (1-F) \cdot f_w(\bar{S}) \cdot \text{MRT} = 1$$

Not surprisingly, in the absence of horizontal transmission $\mu = \bar{\mu} = 0$, a parasitic plasmid cannot satisfy the persistence condition since $F = F_+$ in that case. In the interesting case that $\mu, \bar{\mu} > 0$, we cannot assert that $F = F_+$ since F_+ depends on μ and $\bar{\mu}$. However, it is reasonable to speculate that $F - F_+$ is small. Putting $F = F_+$, (24) for plasmid persistence becomes:

$$\text{MRT} \cdot [F \cdot \bar{\mu}\bar{u} + (1-F) \cdot \mu\bar{w}] > 1 - (1-c)(1-q) = c + q - cq \qquad (26)$$

The left hand side of the inequality gives the number of plasmid transfers made by a single plasmid-bearing cell before being washed out of the chemostat. It must exceed a positive threshold which depends on the cost of carriage c and the probability of miss-segregation q for the plasmid to survive.

Equation (26) indicates that the conjugation terms must exceed a threshold to maintain a parasitic plasmid. In order to see this more clearly, we fix $\alpha = \alpha_+ = 0.1$ and $\beta = \beta_+ = 0.4$, with all other parameters as in Fig. 6.4, except that $\bar{\mu} = \mu \times 10^{-3}$. We then vary μ and plot the resulting stable steady state value of $u_+ + \delta w_+$ in Fig. 6.5. This was done by integrating the differential equations, starting with a tiny inoculum of plasmid-bearing cells, to steady state. If $u_+ + \delta w_+ = 0$, as it does for small μ, that means the stable steady state is the plasmid-free state; if $u_+ + \delta w_+ > 0$, then we are plotting the coexistence steady state value of $u_+ + \delta w_+$. This bifurcation diagram shows that the critical value $\mu_c \approx 6 \times 10^4$ at which the coexistence steady state appears is much larger than our order of magnitude estimate of a biologically reasonable value of $\mu \approx 1$.

We are particularly interested in the case that the plasmid-free organism cannot form a macroscopically significant biofilm, for example, this may mean that it can form only a monolayer (see e.g. Pratt and Kolter (1998) and O'Toole and Kolter (1998)), while the plasmid-bearing organism can form a healthy biofilm. In this case, it is reasonable to assume that the plasmid-bearing organisms sloughing rate does not exceed that of the plasmid-free organism and that its adhesion rate constant is not less than that of the plasmid-free organism:

$$\beta_+ \leq \beta \quad \text{and} \quad \alpha \leq \alpha_+. \qquad (27)$$

Strict inequality is assumed to hold in at least one of these. In this case, as noted in the appendix, the plasmid-bearing organism has a residence time

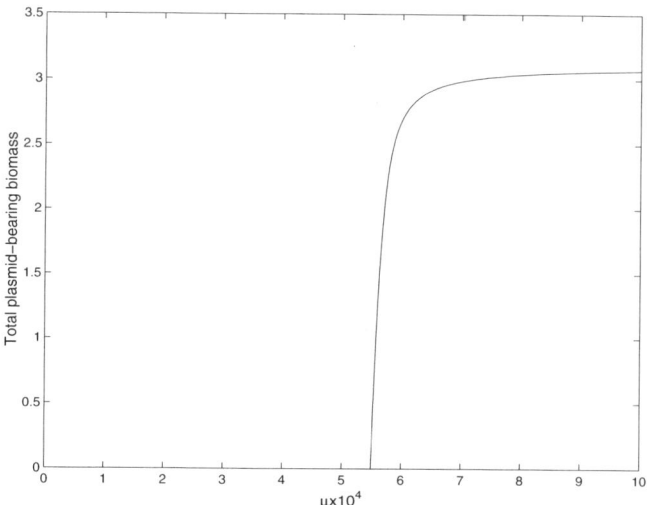

Fig. 6.5. Bifurcation of coexistence steady state from the plasmid-free steady state at a critical value of μ. Vertical axis is the coexistence steady state value of total plasmid-bearing biomass $u_+ + \delta w_+$. Horizontal axis is the conjugation rate μ

advantage over its plasmid-free rival:

$$\text{MRT}_+ > \text{MRT}$$

and it should spend more of this time on the wall than the plasmid-free organism so we conjecture that:

$$F > F_+$$

where, presumably, bacterial densities are higher than in the fluid state and contact rates between organisms are higher:

$$\bar{\mu}\bar{u} >> \mu\bar{w}.$$

(24) says that these advantages must outweigh the cost of carriage and segregational loss.

6.7 Appendix: Mean residence times

Consider the linear compartmental system with no inputs

$$x' = Ax \tag{28}$$

where A is a stable, quasi-positive ($a_{ij} \geq 0, i \neq j$), irreducible $n \times n$ matrix satisfying

$$A = B - C$$

where $C > 0$ and B is quasi-positive with zero column sums. Matrix B accounts for the "internal transitions" of mass between compartments while C accounts for the loss of mass to the external environment. To see this, let $\mathbf{1}$ be the vector of ones. The total mass of x is $\mathbf{1} \cdot x$. Then the rate of loss from the system to the environment is given by

$$(\mathbf{1} \cdot x)' = \mathbf{1} \cdot Ax = \mathbf{1} \cdot Bx - \mathbf{1} \cdot Cx = -\mathbf{1} \cdot Cx \leq 0.$$

Let $x_0 > 0$ be such that $\mathbf{1} \cdot x_0 = 1$. We call x_0 a probability vector. If we start the system off at x_0, then $x(t) = e^{At}x_0$. The probability of still being in the system at time t is given by

$$P(\text{in system at time t}) = \int_t^\infty \mathbf{1} \cdot Cx(s)\,\mathrm{d}s = \mathbf{1} \cdot C \int_t^\infty x(s)\,\mathrm{d}s$$

The mean residence time (MRT) is given by

$$\begin{aligned}
\text{MRT}(x_0) &= \int_0^\infty t\mathbf{1} \cdot Cx(t)\,\mathrm{d}t \\
&= \int_0^\infty C \int_t^\infty x(s)\,\mathrm{d}s\,\mathrm{d}t \cdot \mathbf{1} \\
&= \int_0^\infty C \int_t^\infty e^{As}\,\mathrm{d}s\,\mathrm{d}t\, x_0 \cdot \mathbf{1} \\
&= C \int_0^\infty e^{At} \int_t^\infty e^{A(s-t)}\,\mathrm{d}s\,\mathrm{d}t\, x_0 \cdot \mathbf{1} \\
&= C \int_0^\infty e^{At} \int_0^\infty e^{Ar}\,\mathrm{d}r\,\mathrm{d}t\, x_0 \cdot \mathbf{1} \\
&= C(\int_0^\infty e^{Ar}\,\mathrm{d}r)^2 x_0 \cdot \mathbf{1} \\
&= C(-A^{-1})^2 x_0 \cdot \mathbf{1}
\end{aligned}$$

where we have used that

$$-A^{-1} = \int_0^\infty e^{At}\,\mathrm{d}t\,.$$

We see that the MRT(x_0) depends on x_0, the initial distribution among the compartments. In order to obtain a quantity that is independent of the initial distribution, we might average the above quantity over the standard simplex of probability vectors $\Sigma = \{x_0 \in \mathbb{R}^n : x_0 \geq 0,\ x_0 \cdot \mathbf{1} = 1\}$, obtaining the result

$$\text{MRT} = \frac{1}{|\Sigma|} \int_\Sigma C(-A^{-1})^2 x_0 \cdot \mathbf{1}\,\mathrm{d}S(x_0) = \frac{1}{n} C(-A^{-1})^2 \mathbf{1} \cdot \mathbf{1}$$

where $|\Sigma|$ denotes the area of Σ.

Instead, we note that the matrix A has a dominant eigenvector v corresponding to $s(A) := \max \Re \lambda$ where the maximum is taken over all eigenvalues of A.

$$Av = s(A)v, \quad 1 \cdot v = 1, \quad v > 0.$$

$s(A)$, the stability modulus of A, is an eigenvalue and our assumptions are that $s(A) < 0$. All non-negative solutions of (28) are asymptotic to $x(t) = e^{s(A)t}v$.

If $x_0 = v$, then we may easily compute MRT(v) from the formula above:

$$\text{MRT}(v) = CA^{-2}v \cdot 1 = \frac{1}{s(A)^2} Cv \cdot 1.$$

On the other hand, with $x_0 = v$, $x(t) = e^{s(A)t}v$ so $(1 \cdot x(t))' = s(A)e^{s(A)t}1 \cdot v$. Thus

$$P(\text{leave system at between } t \text{ and } t + dt) = -s(A)e^{s(A)t}1 \cdot v \, dt$$

so we define MRT for (28) to be:

$$\text{MRT} = \int_0^\infty 1 \cdot e^{s(A)t} v \, dt = \int_0^\infty e^{s(A)t} \, dt \, 1 \cdot v = \frac{-1}{s(A)}.$$

As an example consider the case

$$A = A(\alpha, \beta, D) := \begin{pmatrix} -D - \alpha & \beta \\ \alpha & -\beta \end{pmatrix} = \begin{pmatrix} -\alpha & \beta \\ \alpha & -\beta \end{pmatrix} - \begin{pmatrix} D & 0 \\ 0 & 0 \end{pmatrix}$$

A simple calculation gives

$$s(A) = -\left[\frac{D + \alpha + \beta - \sqrt{(D + \alpha + \beta)^2 - 4\beta D}}{2}\right]. \tag{29}$$

The corresponding eigenvector of unit mass is given by

$$v = [-s(A)/D, (1 + s(A)/D)]^\mathrm{T}.$$

Using

$$A^{-1} = \frac{1}{D\beta} \begin{pmatrix} -\beta & -\beta \\ -\alpha & -D - \alpha \end{pmatrix}$$

we find that

$$\text{MRT}(x_0) = \frac{(\beta + \alpha)x_0^1 + (\beta + \alpha + D)x_0^2}{D\beta}.$$

Obviously,
$$\frac{(\beta+\alpha)}{D\beta} \leq \mathrm{MRT}(x_0) \leq \frac{(\beta+\alpha+D)}{D\beta}$$
where the two extremes arise from starting with all cells in the fluid and starting with all cells on the wall. In particular, we obtain the estimate

$$\frac{1}{D} < \frac{(\beta+\alpha)}{D\beta} \leq \mathrm{MRT} = -\frac{1}{s(A)} \leq \frac{(\beta+\alpha+D)}{D\beta}. \qquad (30)$$

Recall that $1/D$ is the mean residence time for a bacterial cell in the fluid for the classical chemostat model. The possibility of a cell adhering to the wall obviously has the consequence that the mean residence time in the chemostat increases. The inequality (30) also implies $v > 0$.

Simple calculations give that

$$\frac{\mathrm{d}}{\mathrm{d}\beta}\mathrm{MRT} < 0, \quad \frac{\mathrm{d}}{\mathrm{d}\alpha}\mathrm{MRT} > 0.$$

It is intuitive that the mean residence time increases with the wall affinity α and decreases with the sloughing rate β.

References

1. Angles, M. and A. Goodman (2000), Plasmid Transfer between Bacteria in Biofilms, in *Biofilms:recent advances in their study and control*, Evans (ed.), (Harwood Academic Publisher, London)
2. Berman, A. and R. Plemmons (1979), *Nonnegative matrices in the mathematical sciences*, (Academic Press, New York)
3. Diekmann, O. and J. Heesterbeek (2000), *Mathematical Epidemiology of Infectious Diseases, Model Building, Analysis and Interpretation*, (Wiley, Chichester).
4. Freter, R. (1983), Mechanisms that control the microflora in the large intestine, in: D.Hentges, (Ed.), *Human Intestinal Microflora in Health and Disease*, (Academic Press, New York).
5. Ghigo, J.-M. (2001), Natural conjugative plasmids induce bacterial biofilm development, Nature 412: 442–445.
6. Imran, M., D. Jones, H.L. Smith, Biofilms and the Plasmid Maintenance question, preprint.
7. Simonsen, L. (1991), The existence conditions for bacterial plasmids: theory and reality, Microbial Ecology 22: 187–205.
8. Bergstrom, C.T., M. Lipsitch, B.R. Levin (2000), Natural selection, infectious transfer and existence conditions for bacterial plasmids, Genetics 155: 1505–1519.
9. Hsu, S.B., P. Waltman, G. Wolkowicz (1994), Global analysis of a model of plasmid-bearing, plasmid-free competition in the chemostat, J. Math. Biol. 32: 731–742.

10. Hsu, S.B. and P. Waltman (1997), Competition between plasmid-bearing and plasmid-free organisms in selective media, Chem. Engng. Sci. 52: 23–35.
11. Hsu, S. B. and P. Waltman (2004), A Survey of Mathematical Models of competition with an Inhibitor, Mathematical Biosciences 187: 53–91.
12. Levin, B.R. and V.A. Rice (1980), The kinetics of transfer of nonconjugative plasmids by mobilizing conjugative factors, Genet. Res. 35: 241–259.
13. O'Toole, G. and R. Kolter (1998), Flagellar and twitching motility are necessary for Pseudomonas aeruginosa biofilm development, Molecular Microbiol. 30: 295–304.
14. Pilyugin, S. and P. Waltman (1999), The simple chemostat with wall growth, SIAM J. Appl. Math. 59: 1552–1572.
15. Pratt, L. and R. Kolter (1998), Genetic analysis of Escherichia coli biofilm formation:roles of flagella, motility, chemotaxis and type I pili, Molecular Microbiol. 30: 285–293.
16. Ryder, D.F. and D. DiBiasio (1984), Biotechnology and Bioengineering, Vol. XXVI: 942–947.
17. Smith, H.L. and P. Waltman (1995), *The Theory of the Chemostat*, (Cambridge University Press, New York).
18. Stemmons, E. and H.L. Smith (2000), Competition in a chemostat with wall attachment, SIAM J. Appl.Math. 61: 567–595.
19. Stephanopoulus, G. and G. Lapidus (1988), Chemostat dynamics of plasmid-bearing plasmid-free mixed recombinant cultures, Chem. Engng. Sci. 43: 49–57.
20. Stewart, F.M. and B.R. Levin (1977), The population biology of bacterial plasmids: a priori conditions for the existence of mobilizable nonconjugative factors, Genetics 87: 209–228.
21. Summers, D. (1996), *The Biology of Plasmids*, (Blackwell Science, London).
22. Thieme, H.R. (1993), Persistence under relaxed point-dissipativity (with application to an epidemic model), SIAM J. Math. Anal. 24: 407–435.
23. Zhao, X.-Q. (2003), *Dynamical Systems in Population Biology*, CMS Books in Mathematics, (Springer, New York).

7

Nonlinearity and Stochasticity in Population Dynamics

J. M. Cushing

Summary. Theoretical studies of population dynamics and ecological interactions tend to focus on asymptotic attractors of mathematical models. Modeling and experimental studies show, however, that even in controlled laboratory conditions the attractors of mathematical models are likely to be insufficient to explain observed temporal patterns in data. Instead, one is more likely to see a collage of many patterns that resemble various dynamics predicted by a deterministic model that arise during randomly occurring temporal episodes. These deterministic "signals" might include patterns characteristic of a model attractor (or several model attractors – even from possibly different deterministic models), transients both near and far from attractors, and/or unstable invariant sets and their stable manifolds. This paper discusses several examples taken from experimental projects in population dynamics that illustrate these and other tenets.

7.1 Introduction

During the last century mathematicians and theoretical ecologists developed a plethora of deterministic models for the dynamics of biological populations and ecological systems. The mathematical analysis of these models, most of which are based on differential or difference equations, is overwhelmingly focussed on the asymptotic dynamics of model solutions. The standard procedure is to locate equilibrium states and perform a linearization stability analysis. In some cases a global analysis of asymptotic dynamics is possible (using Lyapunov functions, Poincaré-Bendixson theory, etc.). Periodic solutions play an important role in some models and their existence and asymptotic stability often preoccupies the mathematician. In more recent years, considerable interest has arisen in more complicated asymptotic dynamics and attractors (such as chaotic attractors), although their study has been mostly by means of computer simulations.

With all the historical and current attention paid to the attractors of deterministic models, one would naturally assume that they must play an important role in our understanding of biological ecosystems and in the description and explanation of observed patterns in population data. Yet, it is widely recognized that there is a serious gap between theoretical models and ecologically data (for example, see (Aber 1997)). Few examples exist of models that provide quantitatively accurate descriptions of population time series data, and even less that provide quantitatively accurate and reliable predictions of population and ecosystem dynamics. Of what use, then, to the ecological sciences – particularly the applied ecological sciences – is the vast literature on mathematical models whose asymptotic dynamics we mathematicians spend so much time and effort analyzing?

What should one expect to see when examining ecological time series data? Should one look for temporal patterns that are explainable by the attractors of deterministic models? Given that "noise" is inevitable in ecological time series data, should one look for "fuzzy versions" of attractors? Something of the sort is usually uttered when noise is mentioned in model studies (although when noise is considered it is usually not carefully modeled). In addition, the (rather obvious) caveat is usually mentioned that too much noise will completely obliterate deterministic attractors (in which case, of course, their role is not clear). To relate a model to data one has to think carefully about the source of the "noise" in the data (i.e., the inevitable deviations of data from model predictions). Are these "errors" due primarily to inaccurate measurements? If so, then of course too much noise will likely obliterate any deterministic trends (attractors or other), and the problem of connecting model to data is more concerned with the problem of obtaining accurate data. Even if data is highly accurate (even exact) there will be deviations of data from model predictions because no model can capture all of the mechanisms that determine the dynamics of a biological population. External forces and internal processes not a part of the model result in "environmental" and "demographic" noise. Another possibility is, then, that one might come to find in an ecological data set that transient dynamics predominate (relative to a given model) and take precedence over model predicted asymptotic attractors. Perhaps it is even the case that ecological data typically exhibit repeated episodes of transients as they are continuously buffeted by stochastic perturbations and, as a result, asymptotic attractors play only a small role or even no role at all.

The answers to these questions can determine what one looks for in data and what tools one uses to analyze data; in other words, they can determine what one actually "observes" in data and hence one's judgement about the "validity" of a model and the accompanying theory.

For ecology to become a more precise science and to raise its principles above qualitative descriptions and general verbal metaphors, it is necessary to make stronger connections between models and data. This involves not just new deterministic model equations and their mathematical analysis, but methods to deal with model parameterization/validation and stochasticity (the inevitable deviation of data from model predictions). A time tested procedure used in science to connect theory and models to data is to isolate phenomena, under controlled and replicated experimental conditions, and to manipulate and perturb a system in order to observe its responses. The understanding resulting from such experimental and modeling procedures form a basis for the study of larger scale systems. To quote E. O. Wilson (2002):

> "When observation and theory collide, scientists turn to carefully designed experiments for resolution. Their motivation is especially high in the case of biological systems, which are typically far too complex to be grasped by observation and theory alone. The best procedure, as in the rest of science is first to simplify the system, then to hold it more or less constant while varying the important parameters one or two at a time to see what happens."

It was in this spirit that I began a collaboration nearly fifteen years ago with a team of mathematicians, statisticians and biologists (R. F. Costantino, R. A. Desharnais, B. Dennis and more recently including S. M. Henson, and A. King). This team's collaborations has had two broad goals. First, we wanted to derive and validate a successful model for the dynamics of an experimental population (in this case, species of *Tribolium*). We sought a model that makes quantitatively accurate descriptions of observed data and that we could show makes accurate predictions, under a wide variety of circumstances – predictions that could be corroborated by means of controlled experiments. Second, we would then use our model/experimental system to conduct studies of a wide range of nonlinear phenomena. Initially our fundamental focus was on the asymptotic dynamics predicted by a deterministic model (although we developed stochastic versions of the model to explain the deviations of data from model predictions in order to validate the model and to conduct simulations). To date, we have successfully used our system (and several adaptations and modifications) to study a long list of dynamic phenomena, including equilibria and periodic cycles, stability and destabilization, bifurcations, quasi-periodic motion, routes-to-chaos, temporal patterns on chaotic attractors, sensitivity to initial conditions, the control of chaos, temporal phase shifting, periodicity due to environmental forcing, nonlinear resonance, multiple attractors, lattice effects, the role of spatial scale on dynamics, the effect of genetic adaptation on population dynamics, and competition between two species. See the books (Caswell 2001) and (Cushing et al. 2003) (and the references cited therein) for expositions of our methods and for many of our results.

The final chapter of the book (Cushing et al. 2003) contains a list of general conclusions concerning the modeling of biological populations and various nonlinear phenomena that we have studied. The purpose of this paper is to elaborate on one of the main conclusions in that list: "full explanation of a ecological times series data is unlikely to be found by analyses that rely solely on deterministic model attractors." Instead, it is suggested that what one is more likely to see in time series data is a mixture – a temporal collage – of many patterns that resemble various deterministic dynamics predicted by a model that arise, perhaps only in part, during randomly occurring temporal episodes. These deterministic "signals" might include one or several attractors, transients both near and far from attractors, and unstable invariant sets and their stable manifolds. Moreover, we found in some of our projects that these deterministic patterns might arise from more than one deterministic model! In this paper, I present several examples taken from our experimental projects that are selected to illustrate these tenets.

7.2 Saddles flybys

In 1980, David Jillson (1980) reported an experiment with *Tribolium castaneum* in which a nonlinear resonance phenomenon was observed in a habitat of periodically fluctuating volume (Henson et al. 1997). Our first example comes from Jillson's control treatments in which the habitat was of constant volume. Figure 7.1 shows plots of the larval stages in the three replicate cultures. Also shown is the model predicted orbit of the LPA model

$$L_{t+1} = bA_t \exp\left(-c_{el}L_t - c_{ea}A_t\right)$$
$$P_{t+1} = (1 - \mu_l) L_t \qquad (1)$$
$$A_{t+1} = P_t \exp\left(-c_{pa}A_t\right) + (1 - \mu_a) A_t$$

with parameter estimates obtain from the data (using maximum likelihood methods and a stochastic version of the model (Dennis et al. 1995; Cushing et al. 1998)). The time unit in this model is two weeks and the generation time is four weeks. The predicted (global) attractor is a 2-cycle. There is also a (unique) positive equilibrium which is a saddle. After a short period of time, two of the three replicate plots of the larval stage resemble the crash-boom cycles predicted by the 2-cycle attractor.

The third replicate is strikingly different, however. Initially it also approaches the 2-cycle attractor, but the approach is interrupted by a long period of subdued oscillation (from $t = 6$ to about $t = 20$ or 21, or in other words over seven generations). Figure 7.1 indicates that the larval stage, during this period, is close to the unstable equilibrium predicted by the model. Figure 7.2 shows the data plotted in three dimensional phase space. The initial approach to the 2-cycle attractor was interrupted by a random event that placed the orbit near the (one dimensional) stable manifold of the saddle equilibrium. The data then closely followed the model predicted stable manifold, until it arrived near the saddle where it lingered for 13 time steps. Subsequently this replicate made an oscillatory departure from the saddle (as predicted by the one dimensional unstable manifold) until it too finally arrived near the 2-cycle attractor.

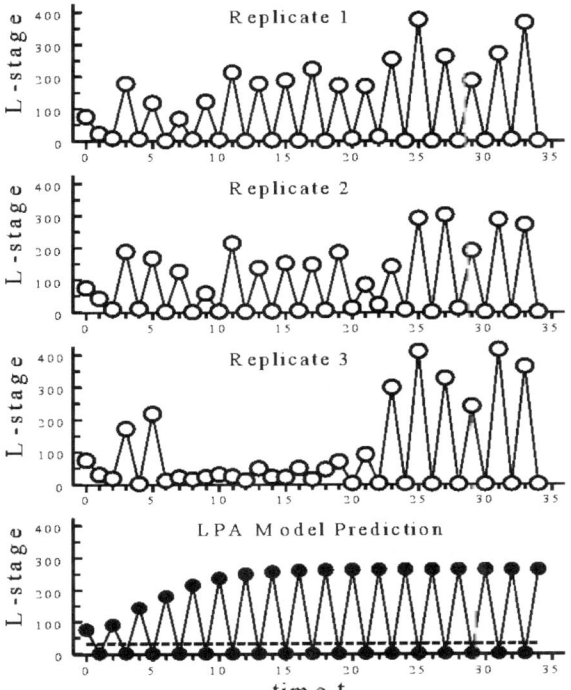

Fig. 7.1. The first three plots show the larval stage of three replicate control cultures from an experiment of Jillson (1980). The fourth plot is that of the LPA model predicted time series of the larval stage with the parameter values $b = 4.44$, $\mu_l = 0.479$, $\mu_a = 0.154$, $c_{el} = 0.0584$, $c_{ea} = 0.00580$, $c_{pa} = 0.0105$. The attractor is a periodic 2-cycle. The dashed line shows the larval component of the model predicted saddle equilibrium

To explain the observed time series in the third replicate of Jillson's controls we see that it is necessary to include not just the model predicted 2-cycle attractor, but also the saddle equilibrium and the geometry of its stable and unstable manifolds. This "unusual" replicate should not be discarded as anomalous (or averaged with the other replicates). Indeed it is valuable. The "saddle flyby" provides more model validation than we would get from time series data that did not visit the saddle (i.e., data orbits like the other two replicates), because it confirms the model predicted dynamics away from the attractor and near the saddle. Stochastic perturbations allow visitation of a wider range of phase space and deepen our understanding of the populations dynamics. (For the same reason they also improve our parameter estimates, since the parameterization procedure is based on the residuals of one-step predictions from each datum point which then have a wider range in phase space (Cushing et al. 2003).)

We have seen such saddle flybys in virtually all of our experimental projects (including saddle cycles as well as saddle equilibria). Figure 7.3 shows another example taken from one of the treatments of a route-to-chaos experiment reported in (Costantino et al. 1997; Cushing et al. 2003; Dennis et al. 2001). In this example, the local unstable manifold is two dimensional (instead of one dimensional as in the example of Fig. 7.1) and is associated with

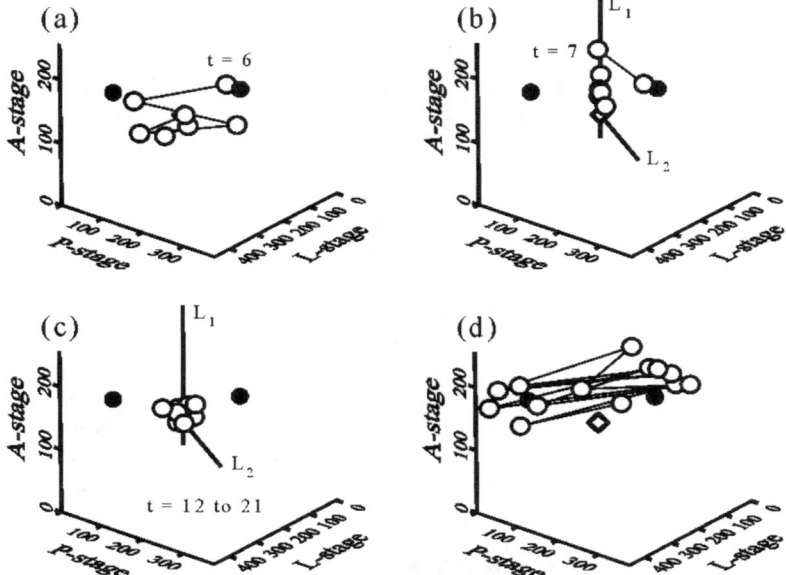

Fig. 7.2. The data from Jillson's replicate 3 produce an orbit in three dimensional phase space. Plots are shown of this orbit over selected temporal subintervals. **a** Initially, from $t = 0$ to 6 the data orbit approaches the 2-cycle attractor denoted by the solid circles. **b** At time $t = 7$ a random perturbation placed the data point near the stable manifold of the saddle equilibrium (denoted by the diamond). The vertical lines L_1 and L_2 are tangents to the two dimensional stable manifold at the saddle (as determined from the eigenvectors of the two eigenvalues $\lambda = 0.80285$ and -0.071169 of the Jacobian matrix respectively). The data orbit from $t = 7$ to 11 closely follow the tangent line L_1. **c** From $t = 12$ to 21 the data orbit lingers near the saddle equilibrium, eventually **d** to return to the 2-cycle attractor

a complex eigenvalue (of magnitude greater than one). The predicted dynamic near the equilibrium is, therefore, quite different from that in Fig. 7.1. The departure of orbits from the unstable equilibrium is expected to be "spiral-like" (with a rotational angle predicted by the argument of the complex eigenvalue). The observed data exhibits this prediction to a remarkable accuracy. This data is from one of three replicates, the other two of which did not undergo such a saddle flyby (Cushing et al. 2003). Notice again that to explain the "anomalous" replicate in Fig. 7.3, as well as the differences between it and the other replicates, we need to include both the attractor and the unstable saddle (and its characteristics) in the analysis.

Sometimes a data time series will undergo a saddle flyby after spending considerable time on or near the attractor. For example, a distinctive saddle equilibrium flyby, lasting 38 weeks (over 9 generations), occurred during the 7th year of an 8 year experiment that placed a culture of *Tribolium* on a chaotic attractor (King et al. 2003). In other examples, saddle flybys oc-

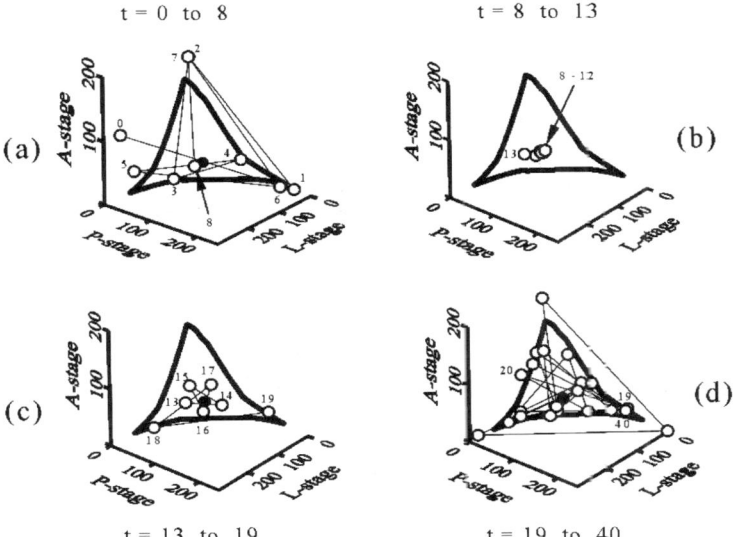

Fig. 7.3. These four graphs show one replicate from one of the treatments of the route-to-chaos experiment reported in (Costantino et al. 1997; Dennis et al. 2001; Cushing et al. 2003). For the estimated and controlled parameter values ($b = 10.45$, $\mu_l = 0.2000$, $\mu_a = 0.9600$, $c_{el} = 0.01731$, $c_{ea} = 0.01310$, $c_{pa} = 0.05000$) the LPA model predicts an invariant loop attractor, appearing in the graphs as a triangular shaped loop. The four graphs show the data orbit broken into four temporal segments. The first and fourth segments in graphs **a** and **d**, corresponding to the beginning and the end of the experiment, show a temporal motion around the model predicted invariant loop. A notable perturbation away from the loop attractor occurs when a stochastic event at $t = 8$ (week 16) placed the data point near a model predicted equilibrium. Graph **b** shows this second segment of the orbit which lingers near the unstable equilibrium for $t = 8$ to 13 (about 8 weeks or, in other words, two generations). The saddle equilibrium has a two dimensional unstable manifold (the linearization has complex eigenvalues of magnitude greater than one) and therefore the model predicts a rotational departure from the equilibrium with, as it turns out, an rotational angle of approximately 145 degrees. This rotation is clearly seen in the data plotted in **c**

cur more than once in a single time series of data; see (Cushing et al. 2003, p. 142) for an example that occurred in the route-to-chaos experiment.

A stochastic version of a deterministic model provides a means by which to study such randomly occurring saddle flybys. We can view simulations of a stochastic model as possible outcomes of an experiment (and repeated simulations as replicates of the experiment). Such a model should not be derive in a cavalier fashion. It is not always appropriate, for example, simply to add noise to the right hand side of the equations in a dynamic model, as is often done. Instead one should place random variables of an appropriate

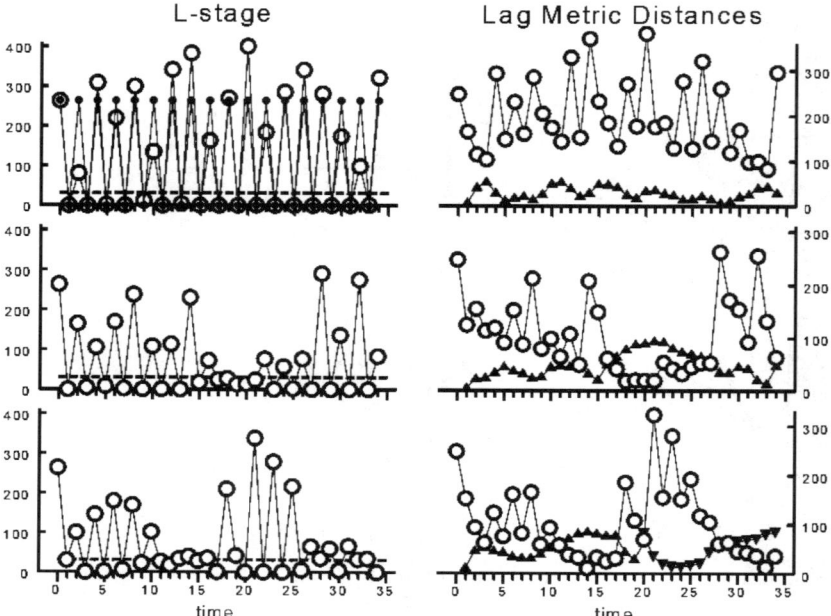

Fig. 7.4. The graphs in the left column show three simulations of a demographic stochasticity version of the LPA model (1) with parameter values as in Fig. 7.1. Noise was added to each of the three equations in the LPA model on the square root scale (uncorrelated normal random variables with variances 10, 1 and 1 respectively) (Dennis et al.1995; Cushing et al. 2003). Simulations were started near the 2-cycle attractor, plotted as the solid circles in the upper graph. The open circles graphs show the L-stage component of the simulations and the dashed line that of the saddle equilibrium. The upper graph show no saddle flyby, while that in the middle graph shows one and the bottom graph shows two flybys. The open circles in the right column graphs show Euclidean distance to the saddle of the simulated orbits at each point in time. The solid triangles show the average of the Euclidean distances of the orbit point and its immediate predecessor from the two points on the 2-cycle attractor. (The triangles pointing up are distances to the phase of the 2-cycle shown in the upper graph in the left column, while the triangles point down are the distances to its phase shift)

kind in appropriate terms, so as to describe the type of stochasticity present in the biological system of interest. Figure 7.4 shows three realizations of a version of the LPA model that approximates demographic stochasticity[1],

---

[1] This model adds a normal random variable of mean zero to each of the three equations in the LPA model on a square root scale. These random variables are uncorrelated in time. In these simulations covariances among them are assumed equal to zero. This kind of stochastic model is one way to describe demographic stochasticity. See (Dennis et al. 1995; Cushing et al. 2003).

with parameter values from Jillson's experiment in Fig. 7.1, that were selected to illustrate saddle flybys. Using a stochastic model, one can study what the model predicts will likely be observed in experimental or observational data (the frequency of flybys, transient characteristics due to the geometry of the saddle in phase space, the relative roles of transients and the attractors, etc.).

While the sorting out of the transient and attractor aspects of time series data not might be difficult in some examples, such as that in Figs. 7.1 and 7.4, in other cases it can fraught with difficulties and pitfalls. If, in an investigation of a data set, one focuses only on attractors and uses diagnostic methods designed for attractors, in a situation when transients are abundant, then obviously it is possible that erroneous conclusions will be drawn. This is particularly true when the attractor is complicated and complex. For example, if stochastically produced transients cause orbits to often revisit the neighborhood of a saddle (or even a repellor), then a large portion of time is spent in regions of phase space where there is exponential separation of orbits. Lyapunov exponents are diagnostic quantities for chaos based on an asymptotic average taken over the attractor. Applying this diagnostic to an orbit that spends enough time near a saddle or repellor can result in the erroneous conclusion that chaos is present. A specific example is given in (Desharnais et al. 1997b), using a stochastic version of the famous Ricker map, in which a "noisy equilibrium" is erroneously diagnosed as chaos by using Lyapunov exponents. Also see (Dennis et al. 2003).

7.3 Basin hopping

Saddles and their stable manifolds also occur as boundaries between basins of attraction in models with multiple attractors. While a deterministic model with multiple attractors makes clear-cut predictions about the asymptotic dynamics of orbits (depending on the initial conditions), when noise is present the dynamics can become complicated, and saddles on the basin boundaries of attraction can play an important role in what dynamic patterns are predicted to be observed in experimental (or simulation) data.

A striking example of this occurs in one of our experiments designed to observed a model predicted, two attractor scenario in a modification of the Jillson experiments (Jillson 1980). Jillson investigated the dynamics of *T. castaneum* in a periodically varying habitat by alternating the volume of flour medium in which populations are cultured. Our analysis of Jillson's data utilizes the LPA model (1) in which habitat volume V is explicitly introduced:

$$L_{t+1} = bA_t \exp\left(-\frac{c_{el}}{V}L_t - \frac{c_{ea}}{V}A_t\right)$$
$$P_{t+1} = (1 - \mu_l) L_t \qquad (2)$$
$$A_{t+1} = P_t \exp\left(-\frac{c_{pa}}{V}A_t\right) + (1 - \mu_a) A_t .$$

The hypothesis that the interaction (cannibalism) coefficients are inversely proportional to habitat size has been experimentally confirmed (Costantino et al. 1998). In a temporally varying habitat, $V = V(t)$ is a function of t; in a periodically varying habitat $V(t)$ is a periodic function of t.

In our multiple attractor experiment the habitat volume was varied periodically with period two and selected amplitudes (Henson et al. 1999). So, in (2) we have $V(t) = 1 + \alpha(-1)^t$ where α is an amplitude and c_{el}, c_{ea}, and c_{pa} are the coefficients in a standardized unit of volume (in our experiments, the volume occupied by 20 grams of flour medium) under constant habitat conditions ($\alpha = 0$).

For parameter values estimated for *T. castaneum* in a constant habitat ($\alpha = 0$) (Costantino et al. 1997) the LPA model (2) predicts a stable 2-cycle attractor. In a periodically varying habitat ($\alpha > 0$) the model predicts two different 2-cycle attractors that perturbed from the two phases of this 2-cycle. (This is true, in fact, in a rather general setting (Henson 2000).) These 2-cycles, while out-of-phase, are not phase shifts of one another and have distinctively different amplitudes; a large amplitude 2-cycle is a called the "resonance" cycle and a small amplitude 2-cycle is called the "attenuant" 2-cycle (Costantino et al. 1998). An unstable (saddle) equilibrium present when $\alpha = 0$ perturbs to a saddle 2-cycle that sits on the basin boundary separating to the regions of attraction for the resonance and attenuant 2-cycles. This multi-attractor scenario occurs for $0 < \alpha < 0.42$. At $\alpha = 0.42$ the attenuant and saddle 2-cycles annihilate one another in a saddle-node bifurcation, leaving a single 2-cycle – the stable resonant cycle.

The experiments reported in (Henson et al. 1999) verified the occurrence of the LPA model's multiple attractor predictions by growing cultures for appropriately selected amplitudes α of flour volume oscillations between 0 and 1. In particular, the presence of the two 2-cycle attractors – resonant and attenuant – was observed in the experimental data at $\alpha = 0.4$. (One reason this is interesting is because the attenuant oscillation was counter-intuitive biologically and seemed not to be a possible dynamic for the beetles.)

However, an interesting and unexpected phenomenon occurred in the multi-attractor experiment. Each replicate culture whose initial conditions were placed in the attenuant 2-cycle's basin of attraction, while clearly exhibiting the features (quantitatively and qualitatively) of the model predicted attenuant 2-cycle early in the experiment, ultimately moved to the basin of the resonant 2-cycle and assumed that attractor's characteristics. No culture in the experiment made the reverse basin migration. The analysis of the experiment presented in (Henson et al. 1999) showed how the saddle cycle and its stable (two dimensional) manifold exhibited a strong influence on the dynamics. Because of stochastic perturbations, the data orbits underwent flybys of the saddle 2-cycle that caused a lingering near that saddle and the basin boundary, which ultimately resulted in a stochastic jump to the resonant 2-cycle basin. These phenomena are in fact predicted by simulations of a stochastic version of the periodic LPA model (2). (Why reverse basin jumps never occur in this case remains an open question.)

The multi-attractor experiment, and the stochastic model used to explain it, show that the predictions of a deterministic model can be altered by noise in important, but predictable and observable ways. In this experiment (and in the stochastic model) one of the two deterministic attractors becomes, in effect, a transient. While the deterministic model helps to explain the results of the experiment, the stochastic version of the model "corrects" (or modifies) the deterministic predictions and provides deeper understanding and insight into the biological system.

Jillson's experiments also included periodic forcing of the habitat volume with other periods. An analysis of the temporal patterns observed in his data, based on the periodically forced LPA model of period 4 and on attractor basin switching and basin boundary saddles, appears in (Henson et al. 2002). In this case, multiple basin switches (back and forth) are observed in some individual time series.

Stochastic attractor basin hopping has also be used as a means to explain phase shifts in oscillatory data time series in non-fluctuating habitats. See (Henson et al. 1998, 2003).

7.4 Lattice effects

The most ambitious experimental project undertaken by our research team during the last decade involved the investigation of a route-to-chaos. This experiment is reported in (Costantino et al. 1997; Dennis et al. 2001) and summarized in our book (Cushing et al. 2003). An analysis of the "chaos" treatment in this experiment not only illustrates the issues described above – the stochastic "dance" of attractors, saddles, and transients – but uncovered some other interesting modeling issues and dynamic phenomena.

In the eight year (96 generations) time series data from the treatment that was designed to corroborate the chaotic attractor predicted by the deterministic LPA model, one can observe a distinctive recursive temporal pattern – a near 11-cycle. An explanation for this dynamic pattern was found when we discovered that there exists an 11-cycle lying on the chaotic attractor that, although a (unstable) saddle cycle, highly influences motion on the attractor. It was surprising to us that such a subtle pattern is discernible in real population data, especially in the presence of chaos and noise[2].

---

[2] Others have also noted transient periodicity in data. Lathrop and Kostelich (1989) found evidence for saddle cycles in a long series of data from the Belousov–Zhabotinshii reaction. So et al. (1998) found evidence for saddle cycles in neuronal electrophysiological recordings. Kendall et al. (1993) and Schaffer et al. (1993) observed similarities between saddle cycles on a chaotic attractor predicted by an epidemiological model and historical measles case-report data.

Furthermore, one finds an even more prominent cyclic pattern – a near 6-cycle pattern – in the times series data. However, it turns out that there is no 6-cycle on the chaotic attractor (or anywhere else in phase space). There is seemingly no explanation possible for this pattern based on the deterministic LPA model. This mystery was solved when, after thinking about the details of the manipulations performed in the experimental protocol, we investigated various "integerized" version of the LPA model. (The experimental data comes in whole numbers, of course, as do individuals in all life stages of the beetle populations.) See for example the model described by Eqs. (4) below. This and other "lattice" models predict, for the initial conditions of the chaos treatment, that the final state of the orbit should be a 6-cycle that is remarkably similar to the pattern observed in the data (Henson et al. 2001)!

On the other hand, a deterministic lattice model cannot predict chaos, since bounded orbits necessarily reach, in finite time, a periodic cycle. Moreover, there are usually more than one "lattice" attractor in such a model. This is true in the lattice LPA model used for the chaos experiment and, as a result, numerous other cyclic patterns might be observable in the data. But what then becomes of chaos? More generally, what roles do the continuous state LPA model and its asymptotic attractors play?

When noise is added to the lattice LPA model we get a stochastic model that predicts the dynamics of the integer value experimental data. Stochasticity continually produces transients on the lattice and these transients, it turns out, resemble the underlying continuous state space attractor (chaotic, in this case). Thus, simulations of a stochastic integerized model predict an episodic interplay of deterministic patterns – attractor, saddles, and transients – from both the deterministic lattice and the deterministic continuous state space model. This phenomenon is illustrated using simpler "toy" models in (Henson et al. 2001; Cushing et al. 2003) and such an example appears in Fig. 7.5. An analysis of the chaos experiment using these notions appears in (King et al. 2003).

Whereas the experiment was designed to put a population into chaotic dynamics – as predicted by the deterministic, continuous state space LPA model – other deterministic patterns are predicted by the lattice LPA model. Specifically the lattice LPA model identified several cycles of various periods as important on the lattice. Stochastic simulations of the lattice LPA model predicted the observed data should contain (randomly occurring) episodes of all these deterministic patterns – and even occasional flybys of the saddle equilibrium (of the deterministic continuous state space model). Indeed, our analysis of the data showed this to be the case; see Fig. 7.6.

In our analysis of the data obtained from the chaos treatment of our experiment, in order to account for the observed temporal patterns it is not sufficient to consider only the asymptotic (chaotic) attractor predicted by the deterministic, continuous state space LPA model. The chaotic attractor does play a role by contributing observable patterns not predicted by the deterministic lattice model, but conversely so also does the deterministic lattice model

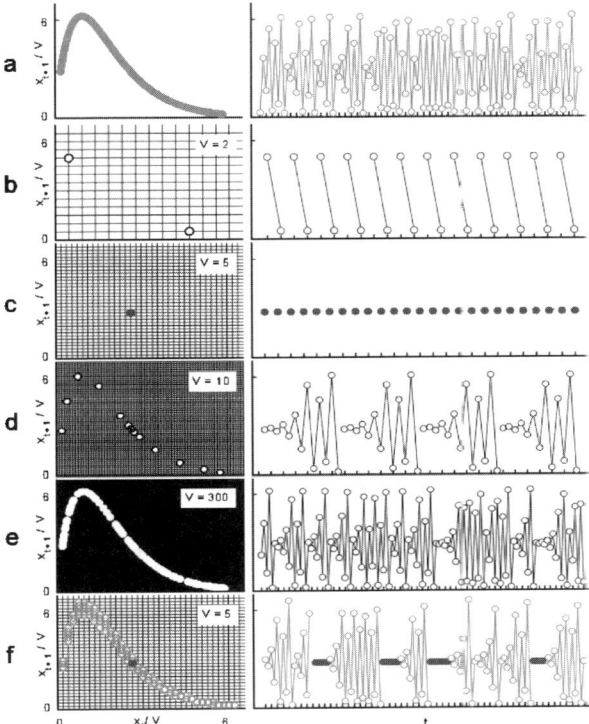

Fig. 7.5. a With $b = 17$ and $c = V = 1$, the Ricker map $x_{t+1} = bx_t \exp(-cx_t/V)$ exhibits chaotic dynamics. **b–e** Show periodic lattice attractors of the integerized Ricker map $x_{t+1} = \text{round}[bx_t \exp(-cx_t/V)]$ with $b = 17$ and $c = 1$ for increasing values of V. Specifically a 2-cycle, 1-cycle (equilibrium), 13-cycle and 117-cycle respectively. In the lagged phase space these attractors (plotted on a density lattice) are seen to increasingly resemble the chaotic attractor. In **f** appears a realization of the (environmental) stochasticity lattice Ricker model $x_{t+1} = \text{round}[bx_t \exp(-cx_t/V) + \sigma z_t]$ with $b = 17$ and $V = c = 1$. Here z_t is a standard normal random variable (uncorrelated in time) and σ measures the magnitude of the noise. This realization is to be compared with the continuous state space, chaotic attractor in **a** and the equilibrium lattice attractor in **c**. Noise has "revealed" the underlying continuous state space chaotic attractor. The time series shows intermittent episodes of both the chaotic and the equilibrium dynamics of the continuous and the lattice models

predict patterns that are not predicted the deterministic chaotic attractor. Stochasticity is needed to explain how these patterns manifest themselves (and in this sense stochasticity becomes an aid and not an obstacle, as it is often viewed).

Henson et al. in (Henson et al. 2003b) consider in more generality the modeling methodology that emerged from the chaos experiment. These authors, using the LPA and other models, discuss how recurrent patterns in stochas-

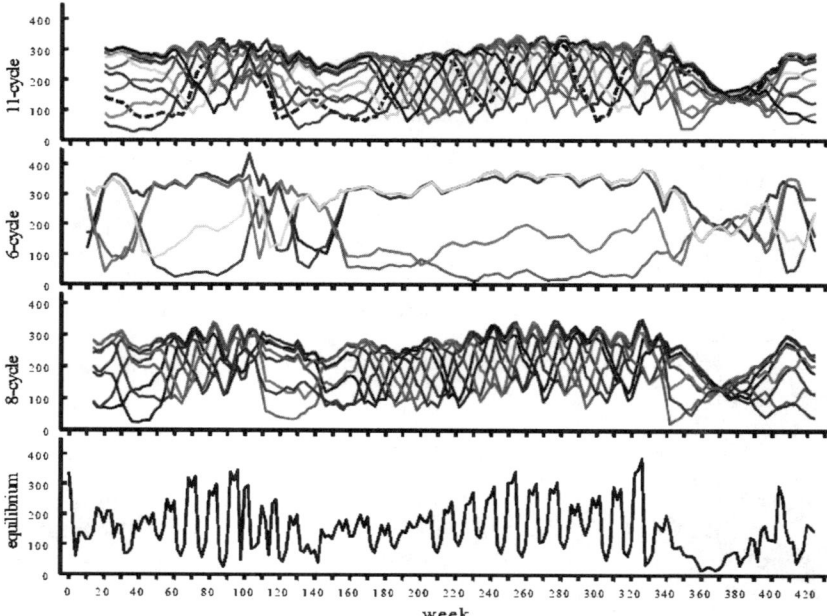

Fig. 7.6. One of the treatments of the route-to-chaos experiment reported in (Costantino et al. 1997; Cushing et al. 2003; Dennis et al. 2001) was based on a chaotic attractor predicted by the LPA model (1) with parameter values $b = 10.45$, $\mu_l = 0.2000$, $\mu_a = 0.9600$, $c_{el} = 0.01731$, $c_{ea} = 0.01310$, $c_{pa} = 0.9600$. A "signature" of the temporal dynamics on the chaotic attractor is a distinctive 11-cycle. The top graph shows the eleven lag metrics (one for each phase of the 11-cycle) computed using one replicate from the experimental treatment. The lag metric measures the average distance of eleven consecutive data points from the corresponding points on a selected phase of the 11-cycle. A low value indicates that the data was close to the 11-cycle for eleven consecutive time steps. The "unravelled" portions of the lag metric braid indicate time intervals during which the data followed closely this signature of the chaotic attractor. (Recall that one generation is 4 weeks.) The LPA model on an integer lattice predicts the experimental initial conditions (and many others) ultimately arrive at a 6-cycle. The graph second from the top shows the lag metrics for the lattice 6-cycle computed from the data. Unravelled portions indicate intervals during which the data was close to this lattice model "attractor". The lattice LPA model has several other cycle attractors, one of which is an 8-cycle whose lag metrics appear in the third graph. The bottom graph displays the lag metric computed with respect to the saddle equilibrium. It clearly indicates a saddle flyby late in the experiment. More details of this "anatomy" of the chaotic attractor appear in (King et al. 2003)

tic processes can be predicted by various deterministic models derived from a parent stochastic mode.

For example, a probabilistic model for *Tribolium* dynamics (based on models of demographic stochasticity in life cycle stage specific birth and death rates) is described by the equations

$$L_{t+1} \sim \text{Poisson}\left[ba_t \exp\left(-\frac{c_{ea}}{V}a_t - \frac{c_{el}}{V}l_t\right)\right]$$
$$P_{t+1} \sim \text{binomial}\left[l_t, 1 - \mu_l\right]$$
$$R_{t+1} \sim \text{binomial}\left[p_t, \exp\left(-\frac{c_{pa}}{V}a_t\right)\right] \qquad (3)$$
$$S_{t+1} \sim \text{binomial}\left[a_t, 1 - \mu_a\right]$$
$$a_{t+1} = r_t + s_t .$$

Here R_t is the number of sexually mature adult recruits, S_t is the number of surviving mature adults, and l_t, p_t, r_t and s_t are the respective numbers observed at time t. The total number of mature adults is $A_t = R_t + S_t$ and $a_t = r_t + s_t$ is the number of mature adults observed at time t. The symbol '$\sim$' means 'is distributed as'. This "Poisson/binomial" LPA (or PBLPA) model is integer value and its dynamics occur on a lattice.

One way to construct a deterministic "skeleton" for the PBLPA model is by iterating the conditional expectation (so that the "most likely" data triple $(L_{t+1}, P_{t+1}, A_{t+1})$ to occur at time $t+1$, given the observed triple (l_t, p_t, a_t), is assumed to be the mean of the random variables in the PBLPA model). This results in the continuous state space LPA model (2).

On the other hand, we can obtain a deterministic skeleton that remains on the integer lattice (where real data is observed) by using another measure of central tendency, namely, the mode. By iterating the conditional mode we obtain a deterministic lattice mode described, as it turns out (assuming the unlikely event of a non-unique conditional mode), by the equations[3]

$$L_{t+1} = \text{floor}\left[bA_t \exp\left(-\frac{c_{ea}}{V}L_t - \frac{c_{ea}}{V}A_t\right)\right]$$
$$P_{t+1} = \text{floor}\left[(1 - \mu_l)(L_t + 1)\right] \qquad (4)$$
$$A_{t+1} = \text{floor}\left[(P_t + 1)\exp\left(-\frac{c_{pa}}{V}A_t\right)\right] + \text{floor}\left[(1 - \mu_a)(A_t + 1)\right] .$$

---

[3] These equations result from formulas for the mode of a binomial random variable and the mode of a Poisson random variable. The following derivations are due to Michael Trosset and Shandelle Henson (private communication). The pdf for a binomial random variable binomial(n, p) is $f(x) = \frac{n!}{x!(n-x)!}p^x(1-p)^{n-x}$. If $x = m$ is the mode, then $f(m+1) \leq f(m)$ and hence $p(n+1) - 1 \leq m$. Also $f(m-1) \leq f(m)$ implies $m \leq p(n+1)$. Since m is an integer, and since $p(n+1)$ is almost always an integer, it follows that $m = \text{floor}[p(n+1)]$. The pdf fo a Poisson random variable poisson(μ) is $f(x) = \frac{\mu^x e^{-\mu}}{x!}$. For the mode m, we see that $f(m+1) \leq f(m)$ implies $\mu - 1 \leq m$ and $f(m-1) \leq f(m)$ implies $m \leq \mu$. Since μ is almost always not an integer, we have $m = \text{floor}[\mu]$. We also point out that the equation for A_{t+1} is different from that given in (Cushing et al. 2003) because of the nature of the experimental protocol involved in the study discussed in that book.

Examples (in addition to the LPA models and the chaos experiment) given by Henson et al. (2003b) show how temporal patterns from both mean (continuous state space) and mode (lattice state space) models are evident in realizations of a stochastic model.

Notice that from this point of view it is not so appropriate to inquire whether or not a specific time series of ecological data has a particular dynamic predicted by a deterministic model, and thus to identify the time series with some type of asymptotic attractor (equilibrium, limit cycle, chaos, etc.). Instead, one expects to observe intermittent episodes of various kinds of patterns, attractor and transient, from perhaps more than one deterministic skeleton. If one expects to see, and only looks for, deterministic attractor patterns, then the modeling exercise used to study the data might be judged a failure when in fact it is very much a success – a success because it can, using an expanded analysis as described above, successfully explain the observed temporal patterns.

For example, suppose one is looking for evidence of chaotic dynamics in time series data. How reliable are conclusions (pro or con) obtained from techniques and diagnostics (e. g., Lyapunov exponents) that are based on the assumption that the data is on an attractor (with some noise, of course), when, in fact, the dynamics might exhibit a stochastic "dance" of attractors, saddles, and transients (Dennis et al. 2003)? A chaotic attractor could be a role player – in this "dance" – and the fact be overlooked. If we found this to be so in the controlled environment and accurately censused populations cultured in our laboratory, then we would expect it to be so, perhaps even more prominently, in field situations.

7.5 Habitat size

Another issue, relating to the important issue of scale in ecology, arose from our route-to-chaos experiment. The predictions of a lattice model can depend significantly on habitat size. This is the case for (3) or (4), whose dynamics change in important ways with the volume V. This is not the case with the continuous state space LPA model (2) whose dynamics only scale with V.

For example, with the estimated and controlled parameter values used in the chaos treatment (Fig. 7.3), a change of V from $V = 1$ (corresponding to the experimental habitat volume occupied by $20\,\mathrm{g}$ of medium) to $V = 3$ ($60\,\mathrm{g}$ of medium) changes the lattice model prediction for the experimental initial conditions from the 6-cycle that played such an important role in the dynamics and analysis at $V = 1$ to a 14-cycle. The 6-cycle is no longer present in the lattice dynamics at the larger habitat volume $V = 3$. Thus, a different collage of patterns would have been predicted and utilized in analysis of the data had the experiment been performed in $60\,\mathrm{g}$ of medium.

In the state space of densities, the number of lattice points increases with V (the lattice mesh size decreases) and the dynamics of the deterministic

7 Nonlinearity and Stochasticity in Population Dynamics   141

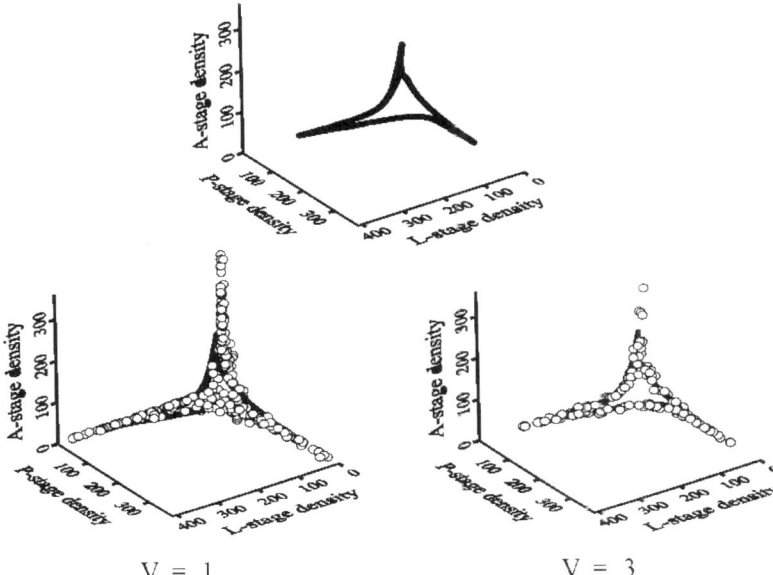

Fig. 7.7. The top graph shows the chaotic attractor, plotted in phase space, predicted by the LPA model in the treatment of the route-to-chaos experiment discussed in the caption of Fig. 7.6. The graph on the lower left shows data points (open circles) from all replicates clustered around the chaotic attractor. This experiment was carried out in a volume occupied by 20 g of standard medium, which corresponds to $V = 1$ in the LPA model (2). A follow-up experiment was conducted in 60 g, or $V = 3$. The results, plotted (as densities) in the lower right hand graph, show a tighter cluster of data points around the chaotic attractor (as predicted by the stochastic lattice model (3))

lattice model converge to the deterministic attractor. This is illustrated for the lattice Ricker model in Fig. 7.5. See Henson et al. (2001, 2003b) for other examples, including the LPA model. Moreover, the stochastic PBLPA model (3) predicts a stronger deterministic (continuous state space) signal as V increases. This is typical of models with demographic stochasticity (May 2001). We have conducted an experiment that duplicates the chaos treatment, but does so in the larger habitat of 60 g ($V = 3$). Although we have not yet published an analysis of this experiment, one can see in Fig. 7.7 that the prediction of a stronger deterministic signal in a larger habitat is supported by the data.

Conversely, the continuous state space attractor is "lost" from the dynamics of the corresponding lattice model if the habitat size is too small. In other words, the size of the habitat effects the predicted dynamic patterns. In the case of chaos, we know of no studies of chaos in ecological data that consider habitat size as a possible factor.

7.6 Concluding remarks

The examples taken from our experimental projects for inclusion in this paper were chosen to illustrate that non-attractor dynamics can play an important role in explaining dynamic patterns observed in data. This is not to say, of course, that attractors are unimportant. Indeed, we designed virtually all of our experimental projects on the basis of model predicted attractors. Nonetheless, we found that in order to obtain a complete and satisfactory explanation of our data it is necessary to include unstable invariant sets, stable manifolds, and so on. This is true even though our experiments involve (seemingly) low dimensional ecosystems cultured in controlled environments in which population counts are highly accurate and stochasticity is minimized. We can successfully account for the dynamic patterns observed in our data by using deterministic model predicted patterns blended together by stochasticity (in most of our cases, demographic stochasticity). In this setting stochasticity becomes an aid, rather than a hindrance, in that it provides the means by which the collage of observed patterns arise (and, in the process, by which the "validation" of the deterministic skeleton that underlies the model is strengthened).

Biological populations and ecosystems are complex, at all levels of organization, and our experience suggests that the mix of stochasticity and nonlinearity will likely be important in most systems. The "higher dimensions" (internal and external) ignored in models with a relatively few number of state variables produces deviations from model predictions (which is modeled as stochasticity). A good example is the plethora of models in which state variables are total population sizes and which in effect treat all individuals as identical, a gross oversimplification in most biological systems. Mathematicians could contribute more to theoretical and applied ecology by extending their efforts beyond the analysis of asymptotic attractors in deterministic models. The study of attractors is, of course, the first step. However, by including stochasticity (in an appropriate way), one can suggest how the deterministic dynamics are likely to manifest themselves in real data. (As we have seen, one can do better than to say that attractors simply made "fuzzy" by noise.) This will strengthen the connection between data and models, and thereby aid ecologists in attempts to account for observed dynamic patterns.

References

1. Aber, J. D. (1997), Why don't we believe the models?, *Bulletin of the Ecological Society of America* 78: 232–23
2. Caswell, H. (2001), *Matrix Population Models: Construction, Analysis and Interpretation*, Second edition, Sinauer Associates, Inc. Publishers, Sunderland, Massachusetts
3. Costantino, R. F., R. A. Desharnais, J. M. Cushing and B. Dennis (1997), Chaotic dynamics in an insect population, *Science* 275: 389–391

4. Costantino, R. F., J. M. Cushing, B. Dennis and R. A. Desharnais and S. M. Henson (1998), Resonant population cycles in temporally fluctuating habitats, *Bulletin of Mathematical Biology* 60: 247–275
5. Cushing, J. M., B. Dennis, R. A. Desharnais, R. F. Costantino (1998), Moving toward an unstable equilibrium: saddle nodes in population systems, *Journal of Animal Ecology* 67: 298–306
6. Cushing, J. M., R. F. Costantino, Brian Dennis, R. A. Desharnais, S. M. Henson (2003), *Chaos in Ecology: Experimental Nonlinear Dynamics*, Theoretical Ecology Series, Academic Press/Elsevier, San Diego
7. Cushing, J. M. (2004), The LPA model, *Fields Institute Communications* 43: 29–55
8. Dennis, B., R. A. Desharnais, J. M. Cushing, and R. F. Costantino (1995), Nonlinear demographic dynamics: mathematical, models, statistical methods, and biological experiments, *Ecological Monographs* 65 (3): 261–281
9. Dennis, B., R. A. Desharnais, J. M. Cushing, S. M. Henson and R. F. Costantino (2001), Estimating chaos and complex dynamics in an insect population, *Ecological Monographs* 71, No. 2: 277–303
10. Dennis, B., R. A. Desharnais, J. M. Cushing, S. M. Henson, R. F. Costantino (2003), Can noise induce Chaos?, *Oikos* 102: 329–340
11. Desharnais, R. A., Costantino, R. F., J. M. Cushing and B. Dennis (June 1997), Letter to editor, *Science* 276: 1881–1882
12. Henson, S. M. (2000), Multiple Attractors and Resonance in Periodically Forced Population Models, *Physica D: Nonlinear Phenomena* 140: 33–49
13. Henson, S. M. and J. M. Cushing (1997), The effect of periodic habitat fluctuations on a nonlinear insect population model, *Journal of Mathematical Biology* 36: 201–22
14. Henson, S. M., J. M. Cushing, R. F. Costantino, B. Dennis and R. A. Desharnais (1998), Phase switching in biological population, *Proceedings of the Royal Society* 265: 2229–2234
15. Henson, S. M., R. F. Costantino, J. M. Cushing, B. Dennis and R. A. Desharnais (1999), Multiple attractors, saddles and population dynamics in periodic habitats, *Bulletin of Mathematical Biology* 61: 1121–1149
16. Henson, S. M., R. F. Costantino, J. M. Cushing, R. A. Desharnais, B. Dennis and Aaron A. King (19 Oct 2001), Lattice effects observed in chaotic dynamics of experimental populations, *Science* 294: 602–605
17. Henson, S. M., R. F. Costantino, R. A. Desharnais, J. M. Cushing, and B. Dennis (2002), Basins of attraction: population dynamics with two stable 4-cycles, *Oikos* 98: 17–24
18. Henson, S. M., J. R. Reilly, S. L. Robertson, M. C. Schu, E. W. Davis and J. M. Cushing (2003a), Predicting irregularities in population cycles, *SIAM Journal on Applied Dynamical Systems* 2, No. 2: 238–253
19. Henson, S. M., A. A. King, R. F. Costantino, J. M. Cushing, B. Dennis and R. A. Desharnais (2003b), Explaining and predicting patterns in stochastic population systems, *Proceedings of the Royal Society London B* 270: 1549–1553
20. Jillson, D. (1980), Insect populations respond to fluctuating environments, *Nature* 288: 699–700
21. Kendall, B. E., W. M. Schaffer and C. W. Tidd (1993), Transient periodicity in chaos, *Physics Letters A.* 177: 13–20

22. King, A. A., R. F. Costantino, J. M. Cushing, S. M. Henson, R. A. Desharnais and B. Dennis (2003), Anatomy of a chaotic attractor: subtle model-predicted patterns revealed in population data, *Proceedings of the National Academy of Sciences* 101, No. 1: 408–413
23. Lathrop, D. P. and E. J. Kostelich (1989), Characterization of an experimental strange attractor by periodic orbits, *Physics Reviews A.* 40: 4028–4031
24. May, R. M. (2001), *Stability and Complexity in Model Ecosystems*, Princeton Landmarks in Biology, Princeton University Press, Princeton, New Jersey
25. Schaffer, W. M., B. E. Kendall, C. W. Tidd and J. F. Olsen (1993), Transient periodicity and episodic predictability in biological dynamics, *IMA Journal of Mathematical Applications to Medicine and Biology* 10: 227–247
26. So, P., J. T. Francis, T. I. Netoff, B. J. Gluckman, and S. J. Schiff (1998), Periodic orbits: a new language for neuronal dynamics, *Biophysics Journal.* 74: 2776–2785
27. Wilson, E. O. (2002), *The Future of Life*, Alfred A. Knopf, New York, p.111

8

The Adaptive Dynamics of Community Structure

Ulf Dieckmann, Åke Brännström,
Reinier HilleRisLambers, and Hiroshi C. Ito

8.1 Introduction

Attempts at comprehending the structures of ecological communities have a long history in biology, reaching right back to the dawn of modern ecology. A seminal debate allegedly occurred between early-twentieth-century plant ecologists Frederic E. Clements and Henry A. Gleason. Textbooks have it (e.g., Calow 1998: 145) that Clements viewed ecological communities as being structured by rich internal dependencies, akin to organisms (Clements 1916), while Gleason held that members of ecological communities were relatively independent of each other, filling ecological niches provided by the abiotic environment (Gleason 1926). While the actual approaches of these two luminaries of plant ecology were more complex than this well-worn caricature suggests (Eliot, in press), their purported positions conveniently established an important conceptual continuum for the mechanistic interpretation of community structures observed in nature.

Modern echoes of this old debate can be found in notions of niche construction (Odling-Smee et al. 2003), leaning towards the Clementsian end of the spectrum, or in the neutral theory of biodiversity and biogeography (Hubbell 2001), which is more in line with a Gleasonian perspective. Like in many other fundamental disputes in ecology, neither side turns out to be simply right or wrong. Instead, disagreements of this kind tend to be resolved at a higher level – by recognizing, firstly, that the original controversy was based on unduly generalized and polarized claims, and secondly, by refocusing scientific attention on elucidating the specific factors and mechanisms that push ecological systems towards either end of the intermediary continuum. Below we will propose such an overarching notion for reinterpreting the Clements–Gleason debate.

Early theoretical models of community structure were based on the simplifying concept of randomly established ecological communities (May 1973). This first wave of models suggested that larger random communities were less likely to possess stable fixed-point equilibria than smaller ones – thus giving rise to yet another long-lasting debate in ecology, about the relationship between community complexity or diversity on the one hand, and community stability or productivity on the other (e. g., Elton 1958; McCann 2000). A second wave of models subsequently imbued such investigations with a higher degree of ecological realism by accounting for the historical route through which new ecological communities are assembled from scratch, and considering more than only infinitely small community perturbations (Post and Pimm 1983; Drake 1990; Law 1999). These assembly models usually relied on the notion of a species pool from which individual species are drawn successively and at random, mimicking the arrival of immigrants from outside an incipient community. A third, much more recent, wave of models rises above considering mere immigration from such a pre-defined species pool, by trying to understand the potential of natural selection for shaping the dynamics and structures of ecological communities (Caldarelli et al. 1998; Drossel al. 2001; Loeuille and Loreau 2005; Ito and Ikegami 2003, 2006). Together, these alternative suites of models suggest that community structures in ecology can only be fully comprehended when processes of interaction (first-wave models), immigration (second-wave models), and adaptation (third-wave models) are taken into account. Appreciating the mechanisms that generate and maintain diversity in ecological communities thus requires methods stretching across the typically different time scales of interactions, immigrations, and adaptations.

Once the dynamics of community formation are recognized to encompass phenotypic adaptation, it is instructive to recast the classic Clements–Gleason debate in terms of fitness landscapes. Under frequency- and density-independent selection, the fitness landscapes experienced by members of an ecological community are independent of the community's composition, directly corresponding to a Gleasonian view. The resultant constant fitness landscapes result in what is known as 'optimizing selection'. By contrast, when the fitness of community members depend on their overall density and individual frequency, fitness landscapes vary with a community's composition. A situation in which this variability is very pronounced, and the frequency- and density-independent components of selection pressures within the community accordingly are relatively weak, neatly corresponds to a Clementsian view. As so often, reality is bound to lie in between these two extremes.

Consequently, an evolutionary perspective on community ecology sheds new light on two fundamental ecological debates. On the one hand, assessing the degree to which fitness landscapes are varying with community composition provides a practical approach for locating specific communities along the Clements–Gleason continuum. On the other hand, evolutionary dynamics literally add new dimensions to concepts of community stability: community structures that are ecologically stable are unlikely also to be evolutionarily stable. This realization challenges earlier conclusions as to how the stability of communities is affected by their complexity or diversity. In particular, ecologically unstable communities may be stabilized by the fine-tuning afforded through coevolutionary adaptations, while ecologically stable communities may be destabilized by evolutionary processes such as arms races, taxon cycles, speciation, and selection-driven extinctions.

In the time-honored quest for understanding community structures, ecology and evolution are thus linked inevitably and intricately, with frequency- and density-dependent selection pressures playing important roles. This sets the stage for considering the utility of adaptive dynamics theory for understanding community structure. Adaptive dynamics theory is a conceptual framework for analyzing the density- and frequency-dependent evolution of quantitative traits, based on a general approach to deriving fitness functions, selection pressures, and evolutionary dynamics from the underlying ecological interactions and population dynamics (e. g., Metz et al. 1992; Dieckmann 1994; Metz et al. 1996; Dieckmann and Law 1996; Geritz et al. 1997, 1998). After introducing the main concepts and models of this approach in Sect. 8.2, this chapter proceeds, in Sects. 8.3 and 8.4, to brief discussions of how selection pressures may drive the increase or decrease, respectively, of species numbers in ecological communities. Armed with this general background, four specific examples of increasingly complex community evolution models are studied in Sects. 8.5 to 8.

Models of evolutionary community assembly are still in their infancy. Accordingly, much room currently exists for investigating systematic variations of already proposed model structures, so as to separate critical from incidental model assumptions and ingredients. The main purpose of this chapter is to introduce readers to a particularly versatile mathematical toolbox for carrying out these much-needed future investigations.

8.2 Models of adaptive dynamics

The theory of adaptive dynamics derives from considering ecological interactions and phenotypic variation at the level of individuals. Extending classical birth and death processes through mutation, adaptive dynamics models keep track, across time, of the phenotypic composition of populations in which trait values of offspring are allowed to differ from those of their parents.

Throughout this chapter we will adhere to the following notation. Time is denoted by t. The number of species in the considered community is N. The values of quantitative traits in species i are denoted by x_i, be they univariate or multivariate. The abundance of individuals with trait value x_i is denoted by $n_i(x_i)$, while n_i denotes the total abundance of individuals in species i. If species i harbors individuals with m_i distinct trait values x_{ik}, its phenotypic density is given by $p_i(x_i) = \sum_{k=1}^{m_i} n_i(x_{ik})\delta(x_i - x_{ik})$, where δ denotes Dirac's delta function. A species with $m_i = 1$ is said to be monomorphic. For small m_i, species i may be characterized as being oligomorphic, when m_i is large, it will be called polymorphic. The community's phenotypic composition is described by $p = (p_1, \ldots, p_N)$. The per capita birth and death rates of individuals with trait value x'_i in a community with phenotypic composition p are denoted by $b_i(x'_i, p)$ and $d_i(x'_i, p)$. Reproduction is clonal, mutant individuals arise with probabilities $\mu_i(x_i)$ per birth event, and their trait values x'_i are drawn from distributions $M_i(x'_i, x_i)$ around parental trait values x_i.

If all species in the community are monomorphic, with resident trait values $x = (x_1, \ldots, x_N)$, and if their ecological dynamics attain an equilibrium attractor, with resident abundances $\bar{n}_i(x)$, the resultant phenotypic composition is denoted by $\bar{p}(x)$. The per capita birth, death, and growth rates of individuals with trait value x'_i will then be given by $\bar{b}_i(x'_i, x) = b_i(x'_i, \bar{p}(x))$, $\bar{d}_i(x'_i, x) = d_i(x'_i, \bar{p}(x))$, and $\bar{f}_i(x'_i, x) = \bar{b}_i(x'_i, x) - \bar{d}_i(x'_i, x)$, respectively. In adaptive dynamics theory, the latter quantity is called invasion fitness. For a mutant x'_i to have a chance of invading a resident community x, its invasion fitness needs to be positive. The notion of invasion fitness $\bar{f}_i(x'_i, x)$ makes explicit that the fitness $\bar{f}_i$ of individuals with trait values x'_i can only be evaluated relative to the environment in which they live, which, in the presence of density- and frequency-dependent selection, depends on x. Invasion fitness can be calculated also for more complicated ecological scenarios, for example, when species exhibit physiological population structure, when they experience non-equilibrium ecological dynamics, or when they are exposed to fluctuating environments (Metz et al. 1992). If a community's ecological dynamics possess several coexisting attractors, invasion fitness will be multi-valued. While strictly monomorphic populations will seldom be found in nature, it turns out that the dynamics of polymorphic populations can often be well approximated and understood in terms of the simpler monomorphic cases. For univariate traits, depicting the sign structure of invasion fitness results in so-called pairwise invasibility plots (Matsuda 1985; van Tienderen and de Jong 1986, Metz et al. 1992, 1996; Kisdi and Meszéna 1993; Geritz et al. 1997).

Derivatives of invasion fitness help to understand the course and outcome of evolution. The selection pressure $g_i(x) = \frac{\partial}{\partial x'_i} \bar{f}_i(x'_i, x)|_{x_i = x'_i}$ acting on trait value x_i is given by the local slope of the fitness landscape $\bar{f}_i(x'_i, x)$ at $x'_i = x_i$. When x_i is multivariate, this derivative is a gradient vector. Selection pressures in multi-species communities are characterized by $g(x) = (g_1(x_1), \ldots, g_N(x_N))$. Trait values x^* at which this selection gradient vanishes, $g(x^*) = 0$, are called evolutionarily singular (Metz et al. 1992). Also the signs of the second derivatives of invasion fitness at evolutionarily singular trait values reveal important information. When the mutant Hessian $h_{mm,i}(x^*) = \frac{\partial^2}{\partial x'^2_i} \bar{f}_i(x'_i, x)|_{x'_i = x^*_i, x = x^*}$ is negative definite, x^*_i is at a fitness maximum, implying (local) evolutionary stability. When $h_{mm,i}(x^*) - h_{rr,i}(x^*)$ is negative definite, where $h_{rr,i}(x^*) = \frac{\partial^2}{\partial x^2_i} \bar{f}_i(x'_i, x)|_{x'_i = x^*_i, x = x^*}$ denotes the resident Hessian, subsequent invasion steps in the vicinity of x^*_i will approach x^*_i, implying (strong) convergence stability.

8 The Adaptive Dynamics of Community Structure

Based on these considerations, four classes of models are used to investigate the adaptive dynamics of ecological communities at different levels of resolution and generality. Details concerning the derivations of these models are provided in the Appendix and their formal relations are summarized in Fig. 8.2. We now introduce these four model classes in turn.

Individual-based birth-death-mutation processes: polymorphic and stochastic. Under the individual-based model specified above, polymorphic distributions of trait values stochastically drift and diffuse through selection and mutation (Dieckmann 1994; Dieckmann et al. 1995). See Fig. 8.1a for an illustration. Using the specification of the birth, death, and mutation processes provided by the functions b_i, d_i, μ_i, and M_i, efficient algorithms for this class of models (Dieckmann 1994) will typically employ Gillespie's minimal process method (Gillespie 1976).

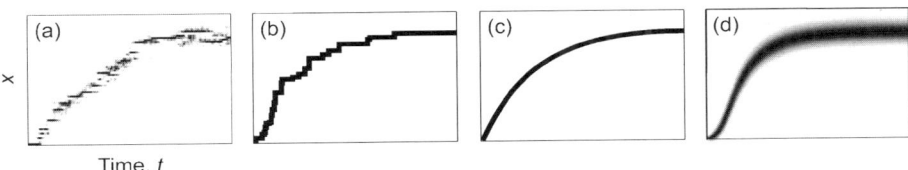

Fig. 8.1. Models of adaptive dynamics. Panel **a** illustrates the individual-based birth-death-mutation process (polymorphic and stochastic), panel **b** shows an evolutionary random walk (monomorphic and stochastic), panel **c** represents the gradient-ascent model (monomorphic and deterministic, described by the canonical equation of adaptive dynamics), and panel **d** depicts an evolutionary reaction-diffusion model (polymorphic and deterministic)

Evolutionary random walks: monomorphic and stochastic. In large populations characterized by low mutation rates, evolution in the individual-based birth-death-mutation process proceeds through sequences of trait substitutions (Metz et al. 1992). During each trait substitution, a mutant with positive invasion fitness quickly invades a resident population, typically ousting the former resident (Geritz et al. 2002). The concatenation of trait substitutions produces the sort of directed random walk depicted in Fig. 8.1b, formally described by the master equation

$$\frac{\mathrm{d}}{\mathrm{d}t}P(x) = \int [r(x,x')P(x') - r(x',x)P(x)]\,\mathrm{d}x'$$

for the probability density $P(x)$ of observing trait value x, with probabilistic transition rates

$$r(x',x) = \sum_{i=1}^{N} \mu_i(x_i)\bar{b}_i(x_i,x)M_i(x'_i,x_i)\bar{n}_i(x)s_i(x'_i,x) \prod_{j=1,j\neq i}^{N} \delta(x'_j - x_j)$$

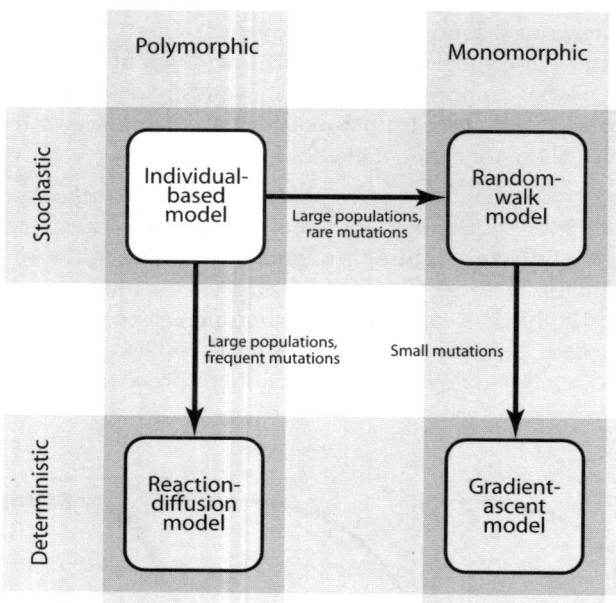

Fig. 8.2. Formal relations between the models of adaptive dynamics. The four classes of model are shown as rounded boxes, and the three derivations as arrows. Arrow labels highlight key assumptions

(Dieckmann 1994; Dieckmann et al. 1995; Dieckmann and Law 1996). Here δ again denotes Dirac's delta function, and $s_i(x'_i, x) = \max(0, \bar{f}_i(x'_i, x))/\bar{b}_i(x'_i, x)$ is the probability with which the mutant x'_i survives accidental extinction through demographic stochasticity while still being rare in the large population of resident individuals (e.g., Athreya and Ney 1972). If also the resident population is small enough to be subject to accidental extinction, $s_i(x'_i, x) = (1 - e^{-2\tilde{f}_i(x'_i, x)})/(1 - e^{-2\tilde{f}_i(x'_i, x)\bar{n}_i(x)})$ with $\tilde{f}_i(x'_i, x) = \bar{f}_i(x'_i, x)/[\bar{b}_i(x'_i, x) + \bar{d}_i(x'_i, x)]$ provides a more accurate approximation (e.g., Crow and Kimura 1970). The resulting evolutionary random walk models are again typically implemented using Gillespie's minimal process method (Dieckmann 1994).

Gradient-ascent models: monomorphic and deterministic. If mutation steps are not only rare but also small, the dynamics of evolutionary random walks are well approximated by smooth trajectories, as shown in Fig. 8.1c. These trajectories represent the evolutionary random walk's expected path and can be approximated by the canonical equation of adaptive dynamics (Dieckmann 1994; Dieckmann et al. 1995; Dieckmann and Law 1996), which, in its simplest form, is given by

$$\frac{d}{dt}x_i = \frac{1}{2}\mu_i(x_i)\bar{n}_i(x)\sigma_i^2(x_i)g_i(x)$$

for $i = 1, \ldots, N$, where

$$\sigma_i^2(x_i) = \int (x'_i - x_i)^T (x'_i - x_i) M_i(x'_i, x_i) \, dx'_i$$

is the variance-covariance matrix of the symmetric mutation distribution M_i around trait value x_i. Implementations of this third class of models typically rely on simple Euler integration or on the fourth-order Runge–Kutta method (e. g., Press et al. 1992).

Reaction-diffusion models: polymorphic and stochastic. In large populations characterized by high mutation rates, stochastic elements in the dynamics of the phenotypic distributions become negligible. This enables formal descriptions of reaction-diffusion type (e. g., Kimura 1965; Bürger 1998). Specifically, the reaction-diffusion approximation of the birth-death-mutation process described above is given by

$$\frac{\mathrm{d}}{\mathrm{d}t}p_i(x_i) = f_i(x_i,p)p_i(x_i) + \frac{1}{2}\sigma_i^2(x_i) * \frac{\partial^2}{\partial x_i^2}\mu_i(x_i)b_i(x_i,p)p_i(x_i)$$

for $i = 1, \ldots, N$, where $\sigma_i^2(x_i)$ is the variance-covariance matrix of the symmetric and homogeneous mutation distribution M_i, and where $*$ denotes the elementwise multiplication of two matrices followed by summation over all resultant matrix elements. An illustration of reaction-diffusion dynamics is shown in Fig. 8.1d. Models of this fourth class are best implemented using so-called implicit integration methods (e. g., Crank 1975). It ought to be highlighted, however, that the infinitely extended tails that the distributions p_i instantaneously acquire in this framework can give rise to artifactual dynamics that offer no good match to the actual dynamics of the underling birth-death-mutation processes in finite populations. The derivation of finite-size corrections to the traditional reaction-diffusion limit overcomes these shortcomings (Dieckmann, unpublished).

At the expense of ignoring genetic intricacies, models of adaptive dynamics are geared to analyzing the evolutionary implications of complex ecological settings. In particular, such models can be used to study all types of density- and frequency-dependent selection, and are equally well geared to describing single-species evolution and multi-species coevolution. As explained above, the four model classes specified in this section are part of a single conceptual and mathematical framework, which implies that switching back and forth between alternative descriptions of any evolutionary dynamics driven by births, deaths, and mutations – as mandated by particular problems in evolutionary ecology – will be entirely straightforward.

8.3 Selection-driven increases in species numbers

Frequency-dependent selection is crucial for understanding how selection pressures can increase the number of species within an ecological community:

- First, whenever selection is optimizing, a single type within each species will be most favored by selection, leaving no room for the stable coexistence of multiple types per species. Frequency-dependent selection pressures, by contrast, can readily create an 'advantage of rarity,' so that multiple types within a species may be stably maintained: as soon as a type's abundance becomes low, the advantage of rarity boosts its growth rate and thus stabilizes the coexistence.

- Second, whereas gradual evolution under optimizing selection can easily bring about stabilizing selection, it can never lead to disruptive selection. This is because, under optimizing selection, the two relevant notions of stability – evolutionary stability on the one hand (Maynard Smith and Price 1973) and convergence stability on the other (Christiansen 1991) – are strictly equivalent: a strategy will be convergence stable if and only if it is evolutionarily stable, and vice versa (e.g., Eshel 1983; Meszéna et al. 2001). Frequency-dependent selection pressures, by contrast, allow for evolutionary branching points, at which directional selection turns disruptive. A gradually evolving population is then trapped at the underlying convergence stable fitness minimum until it splits up into two branches, which subsequently will diverge. This makes the speciation process itself adaptive, and underscores the importance of ecology for understanding speciation.

It is thus clear that frequency-dependent selection is necessary both for the endogenous origin and for the stable maintenance of coexisting types within species.

For univariate traits, the normal form for the invasion fitness of mutants with trait values x' in resident populations with trait values x that are close to an evolutionary branching point with trait value $x^* = 0$ is given by

$$f(x', x) = x'^2 + cx^2 - (1 + c)x'x$$

with $c > 1$ (e.g., Dieckmann 1994: 91). From this we can see that the selection pressure at x^* ceases, $g(x^*) = 0$, that x^* is not locally evolutionarily stable, $h_{mm}(x^*) = 1 > 0$, and that x^* is convergence stable, $h_{mm}(x^*) - h_{rr}(x^*) = 1 - c < 0$. Under these conditions, trait substitutions in x converge to x^* as long as the evolving population is monomorphic, then respond to the disruptive selection at x^* by creating a dimorphism of trait values around x^*, and finally cause the divergence of the two stably coexisting branches away from x^*.

When considering processes of evolutionary branching in sexual populations, selection for reproductive isolation comes into play. As lineage splits are adaptive at evolutionary branching points, in the sense of freeing populations from being stuck at fitness minima, the evolution of premating isolation is favored under such circumstances. Any evolutionarily attainable or already existing mechanism of assortative mating can be recruited by selection to overcome the forces of recombination that otherwise prevent sexual populations from splitting up (e.g., Udovic 1980; Felsenstein 1981). Since there exist a plethora of such mechanisms for assortativeness (based on size, color, pattern, acoustic signals, mating behavior, mating grounds, mating season, the morphology of genital organs etc.), and since only one out of these many mechanisms is needed to take effect, it would indeed be surprising if many natural populations would remain stuck at fitness minima for very long (Geritz et al. 2004). Models for the evolutionary branching of sexual populations corroborate that expectation (e.g., Dieckmann and Doebeli 1999; Doebeli and Dieckmann 2000, 2003, 2005; Geritz and Kisdi 2000; Doebeli et al. 2005).

Processes of adaptive speciation (Dieckmann et al. 2004), resulting from the frequency-dependent mechanisms described above, are very different from those stipulated by the standard model of allopatric speciation through geographical isolation (Mayr 1963, 1982), which have dominated speciation research for decades. Closely related to adaptive speciation are models of sympatric speciation (e.g., Maynard Smith 1966; Johnson et al. 1996), of competitive speciation (Rosenzweig 1978), and of ecological speciation (Schluter 2000), which all point in the same direction: patterns of species diversity are not only shaped by exogenous processes of geographical isolation and immigration, which can be more or less arbitrary, but can instead be driven by endogenous processes of selection and evolution, which are bound to imbue such patterns with a stronger deterministic component.

In conjunction with mounting empirical evidence that rates of race formation and sympatric speciation are potentially quite high, at least under certain conditions (e.g., Bush 1969; Meyer 1993; Schliewen et al. 1994), these considerations suggest that understanding processes and patterns of community formation will crucially benefit from notions developed in the context of adaptive speciation.

8.4 Selection-driven decreases in species numbers

Frequency-dependent selection and density-dependent selection are also crucial for understanding how selection pressures can decrease the number of species within an ecological community:

- First, in evolutionary game theory – including all evolutionary models based on matrix games or on the replicator equation – a population's density is not usually part of the model, which describes only the frequencies of different types. Without enhancements, these types of model therefore cannot account for any density-dependent selection pressures, or capture selection-driven extinctions during which a population's density drops to zero.
- Second, in optimization approaches of evolution, a constant fitness landscape governs the course and outcome of evolution, and, accordingly, frequency-dependent selection is absent. Again, the density of the evolving population is usually not part of the model. Even when it is, selection-driven extinctions cannot occur, as no acceptable constant fitness function will be maximized when a population goes extinct.

These two limitations explain why, until relatively recently, population extinctions caused by natural selection were rarely modeled. In particular, landmarks of evolutionary theory are based on notions of optimizing selection: this includes Fisher's so-called fundamental theorem of natural selection (Fisher 1930) and Wright's notion of hill climbing on fitness landscapes (Wright 1932, 1967). Also Levins's seminal fitness-set approach to the study of constrained bivariate evolution (Levins 1962, 1968) is based on the assumption that, within a set of feasible phenotypes defined by a trade-off, evolution will maximize a population's fitness. Even the advent of evolutionary game theory (Maynard Smith 1982), with its conceptually most valuable refocusing of attention towards frequency-dependent selection, did not help as such, since, for the sake of simplicity, population densities were usually removed from consideration in such models (for an alternative approach to game dynamics aimed at including densities, see Cressman 1990).

And yet the potential of adaptations to cause the collapse of populations was recognized early on. Haldane (1932) provided a verbal example by considering overtopping growth in plants. Taller trees get more sunlight while casting shade onto their neighbors. As selection thus causes the average tree height to increase, fecundity and carrying capacity decline because more of the tree's energy budget is diverted from seed production to wood production, and the age at maturation increases. Arborescent growth as an evolutionary response to selection for competitive ability can therefore cause the decline of a population's abundance as well as of its intrinsic growth rate, potentially resulting in population extinction. The phenomenon of selection-driven extinction is closely related to Hardin's (1968) tragedy of the commons. In both cases, strategies or traits that benefit the selfish interests of individuals, and that are therefore bound to invade, undermine the overall interests of the evolving population as a whole once these strategies or traits have become common. Such a disconnect between individual interest and population interest can only occur under frequency-dependent selection – under optimizing selection, the two are equivalent. It is thus clear that frequency-dependent selection and density-dependent selection are both necessary for capturing the potential of adaptive evolution in a single species to cause its own extinction.

Processes of selection-driven extinction can come in several forms:

- First, evolutionary suicide (Ferrière 2000) is defined as a trait substitution sequence driven by mutation and selection taking a population toward and across a boundary in the population's trait space beyond which the population cannot persist. Once the population's trait values have evolved close enough to this boundary, mutants can invade that are viable as long as the current resident trait value abounds, but that are not viable on their own. When these mutants start to invade the resident population, they initially grow in number. However, once they have become sufficiently abundant, concomitantly reducing the former resident's density, the mutants bring about their own extinction. Webb (2003) refers to such processes of evolutionary suicide as Darwinian extinction.

- Second, adaptation may cause population size to decline gradually through perpetual selection-driven deterioration. Sooner or later, demographic and environmental stochasticity will then cause population extinction. This phenomenon has been dubbed runaway evolution to self-extinction by Matsuda and Abrams (1994a).
- Third, the population collapse abruptly brought about by an invading mutant phenotype may not directly lead to population extinction but only to a substantial reduction in population size (Dercole et al. 2002). Such a collapse will then render the population more susceptible to extinction by stochastic causes and may thus indirectly be responsible for its extinction.

For univariate traits, the normal form for the invasion fitness of mutants with trait values x' in resident populations with trait values x that are close to a critical trait value $x^* = 0$ at which evolutionary suicide occurs is simply given by

$$f(x', x) = x' - x,$$

with the corresponding equilibrium abundance

$$\bar{n}(x) = \begin{cases} 1 + cx^2 & x \leq 0 \\ 0 & x > 0 \end{cases}$$

with $c > 0$. From this we can see that the selection pressure at x^* is positive, $g(x^*) = 1 > 0$, so that trait substitutions in x converge to x^*, where the evolving population's equilibrium abundance abruptly drops from 1 to 0.

The occurrence here of a discontinuous transition to extinction is not accidental. As has been explained by Gyllenberg and Parvinen (2001), Gyllenberg et al. (2002), and Dieckmann and Ferrière (2004), such a catastrophic bifurcation is a strict prerequisite for evolutionary suicide. The reason is that selection pressures at trait values at which a continuous transition to extinction occurs (e. g., through a transcritical bifurcation) always point in the trait direction that increases population size: evolution towards extinction is then impossible. Allee effects, by contrast, provide standard ecological mechanisms for discontinuous transitions to extinction.

The potential ubiquity of selection-driven extinctions is underscored by numerous examples based on the evolutionary dynamics of many different traits, including competitive ability (Matsuda and Abrams 1994a; Gyllenberg and Parvinen 2001; Dercole et al. 2002), anti-predator behavior (Matsuda and Abrams 1994b), sexual traits (Kirkpatrick 1996; Kokko and Brooks 2003), dispersal rates (Gyllenberg et al. 2002), mutualism rates (Ferrière et al. 2002), cannibalistic traits (Dercole and Rinaldi 2002), maturation reaction norms (Ernande et al. 2002), levels of altruism (Le Galliard et al. 2003), and selfing rates (Cheptou 2004). Dieckmann and Ferrière (2004) showed, by examining ecologically explicit multi-locus models, that selection-driven extinction robustly occurs also under sexual inheritance. Relevant overviews of the mathematical and ecological underpinnings of selection-driven extinction have been provided by Webb (2003), Dieckmann and Ferrière (2004), Rankin and López-Sepulcre (2005), and Parvinen (2006).

Also coevolutionary dynamics can cause extinctions. An early treatment, which still excluded the effects of intraspecific frequency-dependent selection, was provided by Roughgarden (1979, 1983). This limitation has been overcome in modern models of coevolutionary dynamics based, for example, on the canonical equation of adaptive dynamics (e.g., Dieckmann et al. 1995, Dieckmann and Law 1996; Law et al. 1997). Also in this multi-species context it is important to distinguish between continuous and discontinuous transitions to extinction. As has been explained above, evolutionary suicide cannot contribute to an evolutionarily driven continuous transition to extinction. Moreover, such continuous extinctions cause mutation-limited phenotypic evolution in the dwindling species to grind to a halt, since fewer and fewer individuals are around to give birth to the mutant phenotypes that fuel the adaptive process. This stagnation renders the threatened species increasingly defenseless by depriving it of the ability to counteract the injurious evolution of its partner through suitable adaptation of its own. For these two reasons, continuous evolutionary extinctions are driven solely by adaptations in the coevolving partners. By contrast, when a transition to extinction is discontinuous, processes of evolutionary suicide and of coevolutionary forcing may conspire to oust a species from the evolving community.

8.5 First example of community evolution: monomorphic and deterministic

Simple community modules comprising two, three, or four interacting species have often been used for investigating how trophic interactions organize simple communities. These studies have laid the foundations for theories (i) of competition, including the R^* rule (Tilman 1982), (ii) of predation within the context of exploitative ecosystems, including work on trophic cascades (Oksanen et al. 1981; Oksanen and Oksanen 2000), and (iii) of omnivory, including research on intraguild predation (Holt and Polis 1997; Diehl and Feissel 2000; Mylius et al. 2001; HilleRisLambers and Dieckmann 2003). All these studies, however, did not account for the potential of evolutionary changes in the ecological interactions between the considered species. Overcoming this restriction is important as patterns of species interactions encountered in nature ought to be interpreted in light of not only ecological stability but also of evolutionary stability.

Here we take a step in this direction by investigating the evolution of feeding preferences within a simple community module. In particular, we examine evolutionary dynamics in simple food webs comprising a basal resource and two antagonistic consumer species, where each consumer is capable of feeding on the resource, on its antagonist, or on a combination of both (HilleRisLambers and Dieckmann 2003). The relative investments into resource or antagonist feeding characterize the consumers' feeding preferences and can evolve subject to a trade-off. In this way, all of the classic three-species community modules – including linear food chains, two consumers sharing a resource, omnivory on the part of one consumer, and mutual intra-guild predation between two consumers – can be realized in the model. By examining how feeding preferences – and thus the trophic linkages between species – evolve, we can chart the possible evolutionary pathways connecting all these classic community modules (HilleRisLambers and Dieckmann, submitted). Since density- and frequency-dependent selection pressures are important for addressing these questions, and since it is desirable to derive the considered evolutionary dynamics from the underlying population dynamics, models of adaptive dynamics provide a useful framework for this kind of analysis.

The abundances n_C and n_D of the two antagonistic consumers change according to Lotka–Volterra dynamics, assuming intrinsic mortalities and linear functional responses. The basal energetic input is provided by a dynamic resource, whose abundance n_R changes according to semichemostat dynamics and consumer feeding. The community's population dynamics are thus given by

$$\frac{d}{dt} n_C = n_C(e_{CR} a_{CR} n_R + e_{CD} a_{CD} n_D - a_{DC} n_D - d_C),$$

$$\frac{d}{dt} n_D = n_D(e_{DR} a_{DR} n_R + e_{DC} a_{DC} n_C - a_{CD} n_C - d_D),$$

$$\frac{d}{dt} n_R = r_R(k_R - n_R) - n_R(a_{CR} n_C + a_{DR} n_D),$$

with attack coefficients a, conversion efficiencies e, and intrinsic mortality rates d. The carrying capacity and intrinsic growth rate of the resource are denoted by k_R and r_R, respectively.

The feeding preferences of the two consumers are affected by a trade-off between the attack coefficients for resource feeding and antagonist feeding,

$$a_{iR} = a_{\max,i} x_i^{s_i}, \quad a_{ij} = a_{\max,i}(1 - x_i)^{s_i},$$

for $i = C, D$ and $j = D, C$, where, for consumer i, $a_{\max,i}$ is the maximal attack coefficient, s_i is the trade-off strength, and the adaptive trait $0 \leq x_i \leq 1$ determines the feeding preference, measured as the relative investment into resource feeding. Intermediate feeding strategies, $0 < x_i < 1$, characterize omnivorous consumers. For $s_i > 1$, generalist feeding strategies ($x_i \approx \frac{1}{2}$) are disfavored compared with specialist feeding strategies ($x_i \approx 0, 1$), resulting in specialist advantage. The reverse is true for $s_i < 1$, which thus corresponds to generalist advantage. On this basis, the canonical equations (Sect. 8.2) for the two adaptive traits x_C and x_D are given by

$$\frac{d}{dt} x_i = \frac{1}{2} \mu_i \sigma_i^2 \bar{n}_i a_{\max,i} s_i [e_{iR} x_i^{s_i - 1} \bar{n}_R - e_{ij}(1 - x_i)^{s_i - 1} \bar{n}_j],$$

for $i = C, D$ and $j = D, C$, where equilibrium abundances are denoted by $\bar{n}$, and μ_i and σ_i^2 are the mutation probability and variance in consumer i.

Coevolutionary dynamics unfold within the constraints of ecological coexistence. HilleRisLambers and Dieckmann (2003) found that, in the model specified above, regions of coexistence open up around linear three-species food chains, $(x_C, x_D) = (0, 1), (1, 0)$, where one consumer is a better antagonist feeder, while the other consumer is a better resource feeder. When the trade-off strengths s_C and s_D are varied together, $s_C = s_D$, two extreme scenarios can be distinguished:

- At strong specialist advantage, linear three-species food chains are evolutionarily stable (in the sense of the corresponding trait combinations being asymptotically stable under the adaptive dynamics described by the two simultaneous canonical equations for x_C and x_D). Under these conditions, selection simplifies community structure by causing the evolution of neighboring trait values towards $(x_C, x_D) = (0, 1), (1, 0)$. This means that the better resource feeder will invest even more into resource feeding, while the better antagonist feeder will invest even more into antagonist feeding, until the evolving three-species food chain has become strictly linear.
- At strong generalist advantage, trait combinations ensuring ecological coexistence are severely limited (HilleRisLambers and Dieckmann 2003). Under these conditions, linear three-species food chains become evolutionarily unstable, and both the better resource feeder and the better antagonist feeder evolve towards generalist strategies, which ultimately results in the exclusion of the former by the latter. Also here the end result is a simplified community structure, in this case given by a simple two-species food chain.

At intermediate trade-off strengths, ecologically feasible communities evolve towards linear two- or three-species food chains, largely depending on the initial feeding preference of the better antagonist feeder.

It must be expected that the trade-offs constraining the attack coefficients of consumer species are not identical, $s_C \neq s_D$. Considering intermediate trade-off strengths, we find that if the better antagonist feeder is more constrained at generalist feeding strategies than the better resource feeder, linear food chains are evolutionarily unstable, and evolutionarily stable food webs exhibiting more complex trophic interactions may be realized. Figure 8.3a shows such a coevolutionary attractor with $(x_C, x_D) \neq (0, 1), (1, 0)$. Figure 8.3a also shows that different coevolutionary attractors may coexist. Depending on the initial food web configuration, coevolution leads to one of the three outcomes depicted in Fig. 8.3b: (i) coexistence between two omnivores, (ii) coexistence between an omnivore and a pure antagonist feeder, or (iii) evolutionary exclusion of the better resource feeder. Which of these occurs is affected largely by the initial feeding preference of the better antagonist feeder and also by the relative scaling of the evolutionary rates in the two

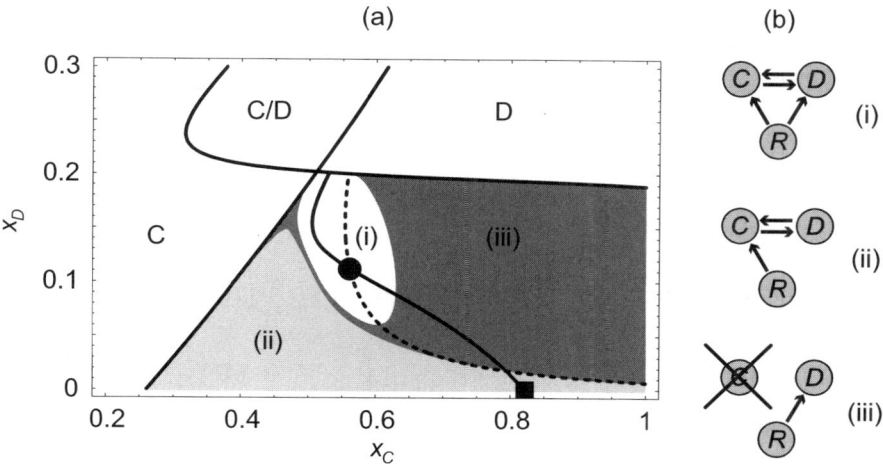

Fig. 8.3. Evolution of community structure in first example. Traits x_C and x_D measure the degree to which consumers C and D invest into feeding on the resource R, as opposed to feeding on each other. For $x_C > x_D$, C is the better resource feeder, while D is the better antagonist feeder. In panel **a**, the evolutionary isoclines of x_C and x_D are depicted by continuous and dashed curves, respectively. Regions in panel **a** indicate different potentials for coexistence and coevolution. Region C: C and R can coexist, while D goes extinct. Region D: D and R can coexist, while C goes extinct. Region C/D: ecological bistability between coexistence of R with either C or D. Regions (i), (ii), and (iii): C D, and R can coexist, so that C and D can coevolve. The community structures resulting from these coevolutionary dynamics then depend on the initial conditions for (x_C, x_D) and are shown in panel **b**. Region (i): Coevolution towards attractor depicted by filled circle, corresponding to omnivorous mutual intraguild predation. Region (ii): Coevolution towards attractor depicted by filled square, corresponding to omnivory on the part of just one consumer. Region (iii): Coevolution towards Region D, corresponding to the exclusion of consumer C. Parameters: $s_C = 0.82$, $s_D = 1.5$, $a_{max,C} = a_{max,D} = 0.4$, $e_{CR} = e_{DR} = 0.2$, $e_{CD} = e_{DC} = 0.8$, $d_C = d_D = 0.05$, $r_R = 0.2$, $k_R = 100$, $\mu_C \sigma_C^2 / \mu_D \sigma_D^2 = 5$

consumers, measured by $\mu_C \sigma_C^2 / \mu_D \sigma_D^2$. Specifically, the basin of attraction for outcome (iii) increases when the better antagonist feeder evolves faster than the better resource feeder. It is also possible that communities of type (i) exhibit cyclical fluctuations in the feeding preferences x_C and x_D, akin to those found by Dieckmann et al. (1995) for predator-prey coevolution and by Law et al. (1997) for coevolution under asymmetric competition. These evolutionary cycles may come dangerously close to the boundaries of coexistence, so that small environmental perturbations may then lead to a shift from outcome (i) to (iii).

We can summarize the results of the analysis here by concluding that linear three-species food chains are most likely to persist evolutionarily under strong specialist advantage, whereas the evolutionary exclusion of consumers is most likely under strong generalist advantage. By contrast, complex trophic interactions in this model are difficult to stabilize evolutionarily. They are most likely to occur in communities in which trade-off strengths are intermediate and the better antagonist feeder experiences a stronger trade-off than the better resource feeder, especially when the latter evolves faster than the former.

8.6 Second example of community evolution: oligomorphic and stochastic

Some existing models of food web evolution incorporate realistic population dynamics, but at the same time rely on interactions mediated by high-dimensional traits that lack clear and direct ecological interpretations (e. g., Caldarelli et al. 1998; Drossel al. 2001). By contrast, a model by Brännström et al. (in preparation), described below, builds on previous foundational work by Loueille and Loreau (2005) and accordingly is based on body size as an evolving trait of high physiological and ecological relevance.

The considered community comprises one autotrophic and N heterotrophic species evolving through mutation-limited phenotypic adaptation. Each species i possesses a trait value x_i determining its body size on a logarithmic scale. From these body sizes, species-specific properties such as energy requirements, competitive interactions, and attack coefficients are determined. The community's demographic processes follow lotka–volterra dynamics, with the dynamics of the non-evolving autotrophic species $i = 0$ given by

$$\frac{\mathrm{d}}{\mathrm{d}t} n_0 = n_0 \left[b_0 - n_0/k_0 - \sum_{j=1}^{N} \exp(x_j - x_0) F(x_j - x_0) n_j \right]$$

and the per capita birth and death rates, respectively, of the heterotrophic species $i = 1, \ldots, N$ given by

$$b_i(x, n) = e \sum_{j=0}^{N} \exp(x_j - x_i) F(x_i - x_j) n_j ,$$

$$d_i(x, n) = d(x_i) + \sum_{j=1}^{N} F(x_j - x_i) n_j + \sum_{j=1}^{N} C(x_i - x_j) \exp(x_j) n_j .$$

The four terms on the two right-hand sides above correspond, in turn, to reproduction, intrinsic mortality, mortality from predation, and mortality from interference competition:

- Energy inflow from foraging results in reproduction as described by the first term. The rate at which new individuals enter the focal species through birth thus depends on the abundance of available prey, on the relative difference in size between predator and prey, and on a predator's ability to attack a prey. The latter is characterized by a shifted Gaussian function F of the relative size difference, with F being referred to as the foraging kernel. The degree to which energy is lost as prey biomass is converted into offspring is measured by the trophic efficiency e.

- The intrinsic mortality rate in the second term is assumed to decrease with body size according to a power law resulting in body-size-dependent generation times consistent with empirical observations (e. g., Peters 1983).
- Losses resulting from predation are captured by the third term, which immediately follows from the considerations concerning the foraging kernel.
- Interference competition between individuals is described by the fourth term. The increase in mortality caused by interference from other individuals depends on their biomass and on the relative size difference. This is characterized by a Gaussian function C, centered at zero and referred to as the competition kernel. Accordingly, two individuals that greatly differ in size will compete much less than two individuals that have similar sizes. The exponential term ensures that smaller individuals are affected more by interference competition.

The evolutionary dynamics of this community are modeled under the assumption that mutations are rare, so that a new mutant will either successfully invade the resident community or be extinct by the time the next successful mutation occurs. We can then employ an oligomorphic extension of the evolutionary random walk model described in Sect. 8.2. Mutations occur at a rate proportional to the total birth rate of the corresponding resident species, and mutant trait values are assumed to be normally distributed around those of their parent. Whether or not a mutant morph can invade the resident community will depend on its invasion fitness, with the success probabilities of potentially invading mutants given in Sect. 8.2. When a successful invasion occurs, its community-level consequences can be determined from the Lotka–Volterra dynamics specified above. However, since the underlying time integration is time-consuming, an approximate, but in practice accurate, algorithm is used, known as the oligomorphic stochastic model (Ito and Dieckmann, unpublished). The steps in this algorithm aim at inferring the structure of the post-invasion community without time integration whenever possible. Simulations of the evolutionary process end when the community-level probability of successful invasion falls below a prescribed threshold.

Figure 8.4a shows how the interplay between mutation and selection gradually leads from a single ancestral species to a community of seven heterotrophic species, through a process of sequential evolutionary branching. The structure of the resulting food web is depicted in Fig. 8.4b.

To isolate and determine the factors governing diversity, two complementary approaches were used. First, the asymptotic number of species was evaluated numerically, as described above. Second, the strengths of disruptive

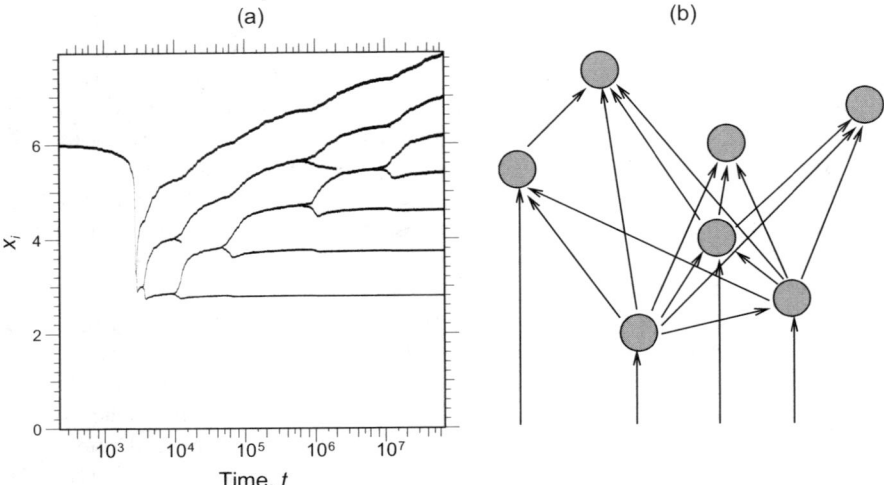

Fig. 8.4. Evolution of community structure in second example. Panel **a** shows the temporal development of community structure through recurrent evolutionary branching, utilizing a logarithmic time scale. Panel **b** depicts the resulting community structure. Each species is represented by a circle, with its vertical position given by its trophic level. Circles are connected by arrows, from prey to predator, where the energy flow between the two corresponding species account for more than 10% of the total energy inflow to the recipient species. Arrows connecting to the bottom indicate consumption of the autotrophic species (or basal resource, which is not displayed). Parameters: $x_0 = 1$, $k_0 = 100$, $b_0 = 1$, $e = 0.3$, $d(x_i) = d_0 \exp(-qx_i)$ with $d_0 = 0.2$ and $q = 0.75$; F is a lognormal function with mean 3, standard deviation 1.5, and amplitude 2.5; C is a lognormal function with mean 0, standard deviation 0.6, and amplitude 0.0025

selection at the first and second branching points were determined as a function of model parameters. This enabled analytical insights into which parameters are important for the initial stages of food web evolution. Interference competition and metabolic scaling (in the form of reduced mortality at larger body size) proved to be critical components in this regard. The former promotes evolutionary branching and is a prerequisite for diversity to develop, while the latter offsets the advantage that smaller species enjoy in terms of increased encounter rates per unit of biomass. In simulations in which either interference competition or metabolic scaling were absent, evolution did not lead to communities with more than just one or two species.

It proved useful to group parameters according to their role in the model, with energy parameters directly affecting the energy flow, foraging parameters determining the shape of the foraging kernel, and competition parameters governing the interference competition between individuals of similar size. With this grouping and terminology in place, it turned out that the initial stages of food web evolution primarily depend on the energy and competition parameters. While these same parameters were naturally also important for the asymptotically evolving diversity, their role there was largely overshadowed by the foraging parameters. The fact that some parameters are mainly important in the early stages of community evolution while others become crucial only during the later stages shows that an analysis that stopped prematurely after investigating only the first or second incidence of evolutionary branching would be insufficient for determining which mechanisms and parameters affect the longer-term structuring of ecological communities.

8.7 Third example of community evolution: polymorphic and deterministic

Explaining the evolutionary origin and history of food webs through sequential adaptive diversification is a challenge that has as yet been tackled by few evolutionary models. It is therefore interesting to explore to what extent the coevolution of predator-prey interactions underlying trophic community structures can induce recurrent evolutionary branching.

In nature the ecological dynamics of phenotypes engaged in trophic interactions depend on how the considered individuals perform in their roles as predator on the one hand and as prey on the other. Both of these components must be expected to evolve. Ito and Ikegami (2003, 2006) therefore considered bivariate adaptive traits $x = (x_r, x_u)$. with the first trait component x_r determining how an individual is exposed as a resource (strategy as prey) and the second trait component x_u determining how the individual is utilizing such resources (strategy as predator). Resources may have many relevant phenotypic properties – including body size, toxicity, proportion of protective tissue, ability to hide, running speed etc. – which jointly can be described by a vector z. The contribution an individual with resource trait x_r makes to the density in this potentially multivariate resource space is denoted by $p_r(x_r, z)$, and analogously the utilization spectrum of an individual with utilization trait x_u is $p(x_u, z)$. Given a phenotypic distribution $p(x)$, the distribution of resource properties is thus $P_r(z) = \int \int p(x_r, x_u) dx_u p_r(x_r, z) dx_r + S(z)$, where S accounts for sources of resource supply from outside the modeled population. Likewise, the population's utilization spectrum is $P_u(z) = \int \int p(x_r, x_u) dx_r p_u(x_u, z) dx_u$. Ito and Ikegami (2003, 2006) then considered the following ecological and evolutionary dynamics,

$$\frac{\mathrm{d}}{\mathrm{d}t} p(x) = \left(e \int F(z) p_u(x_u, z) \,\mathrm{d}z - \int F(z) \frac{P_u(z)}{P_r(z)} p_r(x_r, z) \,\mathrm{d}z - d \right) p(x) + \frac{1}{2} \mu \sigma^2 * \frac{\partial^2}{\partial x^2} p(x) \,,$$

where e measures trophic conversion efficiency and d is the intrinsic death rate. The function $F(z) = aP_\mathrm{r}(z)/(1 + P_\mathrm{r}(z)/P_{1/2})$ is a Holling type II functional response, with maximum a and half-saturation constant $P_{1/2}$. As explained in Sect. 8.2 and in the Appendix, the population-level effect of frequent mutations can be approximated by a diffusion term with diffusion coefficient matrix $\frac{1}{2}\mu\sigma^2$ (to avoid dynamical artifacts, values of $p(x)$ are reset to zero after falling below a very low cutoff threshold).

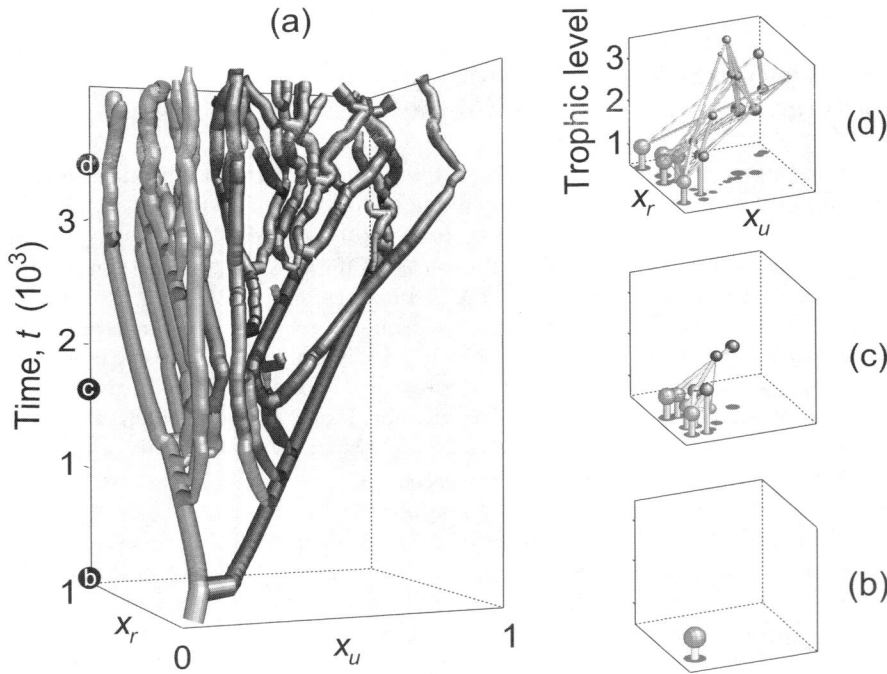

Fig. 8.5. Evolution of community structure in third example. When two trait components for an individual's strategy as prey, x_r, and for its strategy as predator, x_u, evolve under selection pressures resulting from predator-prey interactions, complex food webs can emerge through recurrent evolutionary branching. Panel **a** shows the temporal development of community structure, with the widths of tubes reflecting the densities of phenotypic clusters. Panels **b** to **d** show the evolving food web at three different moments in time. Spheres represent phenotypic clusters, with diameters reflecting the corresponding densities. On the bottom planes, the shadows of these spheres show the distribution $p(x)$. Tubes represent trophic links, with diameters reflecting the corresponding interaction strengths. Tubes connecting to the bottom planes indicate consumption of the external supply of resources (which is assigned trophic level 0). The resultant trophic levels of phenotypic clusters are shown along the vertical axes in **b** to **d**. Parameters: $e = 0.1$, $d = 1$, $a = 20$, $P_{1/2} = 17$, $\frac{1}{2}\mu\sigma^2 = ((3 \cdot 10^{-2}, 0), (0, 10^{-3}))$, $S_0 = 200$, $\sigma_S = 0.08$

For the sake of simplicity, here we assume a one-dimensional resource space, strictly localized functions, $p_r(x_r, z) = \delta(z - x_r)$ and $p_u(x_u, z) = \delta(z - x_u)$, where δ denotes Dirac's delta function, a normally distributed source of external resources, $S(z) = S_0 N_{0,\sigma_S^2}(z)$, and traits x_r and x_u confined to the unit interval. Within a wide range of parameter values, the dynamics of initially unimodal phenotypic distributions $p(x)$ then comprises phases of directional evolution and evolutionary branching. Phenotypic clusters with few prey and many predators go extinct, while phenotypic clusters with many prey and few predators rapidly increase in density and subsequently split through evolutionary branching. Since branching in x_r often induces branching in x_u, and vice versa, the branching sequences resulting from this positive feedback bring about a richly structured food web. Large food webs are maintained through a dynamic balance between selection-driven branching and extinction.

Implementation of sexual reproduction, akin to the model by Drossel and McKane (2000), does not change these dynamics qualitatively (apart from the fact that phenotypic clusters become reproductively isolated). Giving p_r and p_u a certain width, by assuming Gaussian functions instead of delta functions, also does not qualitatively affect evolutionary outcomes. Finally, interference competition among predators can be considered by using $F(z) = a P_r(z) / (P_u(z) + P_r(z)/P_{1/2})$, which gives rise to a ratio-dependent functional response (Arditi and Ginzburg 1989) and facilitates the evolutionary origin and maintenance of complex food webs, as illustrated in Fig. 8.5.

8.8 Fourth example of community evolution: polymorphic and stochastic

The examples presented so far may create the impression that trophic interactions were a necessary prerequisite for the evolutionary origin and maintenance of complex community structures. This is clearly not the case. Purely competitive interactions have long been shown to ensure the maintenance of large species numbers, with early work on the species packing problem dating back to MacArthur and Levins (1967), Vandermeer (1970), May (1973), and Roughgarden (1974).

To illustrate and underscore the potential of purely competitive interactions to bring about and structure multi-species communities through evolutionary dynamics including adaptive radiations, we consider adaptations under asymmetric competition. Specifically, we assume that interactions between individuals are affected by a univariate quantitative trait x, of which we may think, for example, as representing stem height in plants or adult body size in animals. In either case, individuals with a small trait value will suffer a lot from competition against individuals with a large trait value, while the reverse effects will often be negligible. And if individuals are too far apart in their trait values, so as to occupy essentially different ecological niches, they will hardly interact at all. These qualitative dependencies are captured by the function

$$C(x - x') = \exp\left(\frac{1}{2}\sigma_C^2\beta^2\right)\exp\left(-\frac{(x - x' + \sigma_C^2\beta)^2}{2\sigma_C^2}\right),$$

which has been used to describe the strength of competition exerted by an individual with trait value x' on an individual with trait value x (Rummel and Roughgarden 1985; Taper and Case 1992). Here $\beta = 0$ corresponds to symmetric competition, while $\beta > 0$ causes asymmetric competition favoring larger trait values. We also assume that trait values differ in their intrinsic carrying capacity,

$$K(x) = K_0 \exp\left(-\frac{1}{2}(x - x_0)^2/\sigma_K^2\right),$$

which, by itself, causes stabilizing selection towards $x = x_0$. On this basis, we can specify the per capita birth and death rates of individuals with trait values x in a community with phenotypic density p,

$$b(x, p) = b_0, \quad d(x, p) = \frac{1}{K(x)}\int C(x - x')p(x')\,\mathrm{d}x' = \frac{1}{K(x)}\sum_{k=1}^{n} C(x - x_k),$$

resulting in simple population dynamics of Lotka–Volterra type.

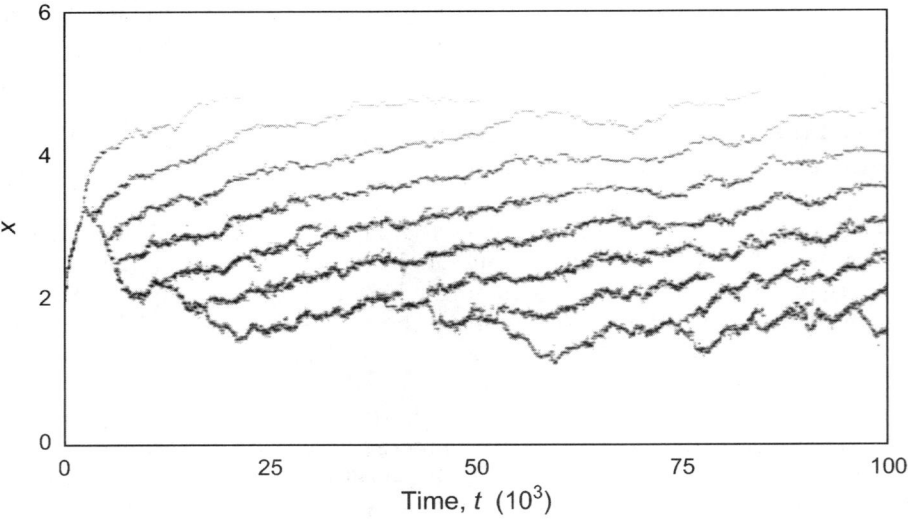

Fig. 8.6. Evolution of community structure in fourth example. When a trait governing asymmetric competition evolves, selection-driven increases and decreases in morph number are embedded into a macroevolutionary pattern of perpetual laminar flow of morphs towards larger trait values. The individual-based dynamics shown involved more than 420,000,000 explicitly simulated birth and death events. Parameters: $b_0 = 1$, $K_0 = 1000$, $x_0 = 2$, $\sigma_K = 1$, $\sigma_C = 0.3$, $\beta = 2$, $\mu = 0.005$, $\sigma = 0.025$

The individual-based birth-death-mutation model introduced in Sect. 8.2 is well suited to explore the resultant evolutionary dynamics (Doebeli and Dieckmann 2000). Figure 8.6 shows a typical realization of this stochastic process. As we can see, directional selection towards larger trait values initially causes convergence to a primary evolutionary branching point. Evolutionary branching subsequently enlarges the number of morphs in the community, until the maximum number resulting from limitations on species packing has been reached. Perpetual coevolutionary change then ensues, through (i) the extinction of morphs with large trait values, which run out of carrying capacity, (ii) the laminar and gradual flow of resident morphs towards the larger trait values favored by asymmetric competition, and (iii) the continual replenishment of morphs at low trait values through adaptive radiations triggered by the opening up of ecological opportunities resulting from the first two effects. It is worthwhile to highlight that in this model the incessant coevolutionary turnover is caused entirely by intrinsic or endogenous mechanisms. No environmental forcing needs to be invoked for understanding the systematic trends in the observed macroevolutionary pattern.

8.9 Summary

In this chapter we have shown how models of adaptive dynamics provide a variety of flexible tools for studying the evolutionary dynamics of ecological communities. Once demography and mutations have been specified, evolutionary and coevolution processes – including those that increase or decrease the number of species in the community – can be analyzed at several mutually illuminating levels of description. While individual-based descriptions of birth, deaths, and mutations provide the finest level of detail, such models are often too computationally intensive and too unwieldy to be comprehensively examined. It is then helpful to have available other classes of models that provide tried and tested approximations. Depending on the features of the evolving community and the nature of the addressed research questions, evolutionary random walks, gradient-ascent models, or reaction-diffusion models may alternatively be best suited for systematically investigating evolving community structures.

Until relatively recently, community models have focused either on the ecological dynamics of large communities or on the evolutionary dynamics of small communities. Now the time seems ripe to bring together these two previously independent strands of inquiry in a new, more ambitious synthesis. Even though it thus has already become clear that a rich diversity of ecological mechanisms can drive the persistent diversification of ecologically relevant adaptive traits, and thus of ecological community structure, much research remains to be done in this area. The eventual goal will be to arrive at a systematic understanding of the ways through which processes of interaction, immigration, and adaptation can work together to generate the rich, but at the same time not arbitrary, community structures observed in nature.

Theoretical models of community evolution are revealing the stunning capacity of ecological interactions, in conjunction with the selection pressures thus engendered, to result in the emergence of non-random community patterns. It thus seems safe to conclude that neither of the old Clementsian or Gleasonian notions – viewing ecological communities as either organismically or externally structured – can do justice to the subtle interplay of endogenous and exogenous demographic and evolutionary pressures unfolding in real communities. Fueled by the mutual shaping and reshaping of ecological niches caused by community evolution, natural community structures appear to occupy a highly complex middle ground.

Appendix: Specification and derivation of adaptive dynamics models

This appendix provides salient mathematical details on how the four models of adaptive dynamics are defined and derived.

Polymorphic Stochastic Model. We start from an individual-based description of the ecology of an evolving multi-species community (Dieckmann 1994; Dieckmann et al. 1995). The number of species in the considered community is N. The phenotypic distribution p_i of a population of n_i individuals in species i is given by $p_i = \sum_{k=1}^{n_i} \delta_{x_{ik}}$, where x_{ik} are the trait values of individual k in species i, and $\delta_{x_{ik}}$ denotes the Dirac delta function peaked at x_{ik}, $\delta_{x_{ik}}(x_i) = \delta(x_i - x_{ik})$. As a reminder we mention that Dirac's delta function is defined algebraically through its so-called sifting property, $\int F(x_i)\delta(x_i - x_0)\,dx_i = F(x_0)$ for any continuous function F. This implies $p_i(x_i) = 0$ unless x_i is represented in species i. We can thus think of $p_i(x_i)$ as a density distribution in the trait space of species i, with one peak positioned at the trait value of each individual in that species. Since $\int \delta_{x_{ik}}(x)\,dx = 1$ for any x_{ik}, we also have $\int p_i(x_i)\,dx_i = n_i$. If $p_i(x_i) \neq 0$ for more than one x_i, the population in species i is called polymorphic, otherwise it is referred to as being monomorphic. The community's phenotypic composition is described by $p = (p_1, \ldots, p_N)$.

The birth and death rates of an individual with trait value x_i in species i are given by $b_i(x_i, p)$ and $d_i(x_i, p)$. Each birth by a parent with trait value x_i gives rise, with probability $\mu_i(x_i)$, to mutant offspring with a trait value $x'_i \neq x_i$, distributed according to $M_i(x'_i, x_i)$, whereas with probability $1 - \mu_i(x_i)$ trait values are inherited faithfully from parent to offspring. A master equation (e.g., van Kampen 1981) describes the resultant birth-death-mutation process,

$$\frac{d}{dt}P(p) = \int [r(p,p')P(p') - r(p',p)P(p)]\,dp'\,.$$

The equation describes changes in the probability $P(p)$ for the evolving community to be in state p. This probability increases with transitions from states $p' \neq p$ to p (first term) and decreases with transitions away from p (second term). A birth event in species i causes a single Dirac delta function, peaked at the trait value x_i of the new individual, to be added to p_i, $p \to p' = p + u_i \delta_{x_i}$, where the elements of the unit vector u_i are given by Kronecker delta symbols, $u_i = (\delta_{1i}, \ldots, \delta_{Ni})$. Analogously, a death event in species i corresponds to subtracting a Dirac delta function from p, $p \to p' = p - u_i \delta_{x_i}$.

The rate $r(p', p)$ for the transition $p \to p'$ is thus given by

$$r(p', p) = \sum_{i=1}^{N} \int [r_i^+(x_i, p) \Delta(p + u_i \delta_{x_i} - p') + r_i^-(x_i, p) \Delta(p - u_i \delta_{x_i} - p')] \, dx_i \,.$$

Here Δ denotes the generalized delta function introduced by Dieckmann (1994), which extends the sifting property of Dirac's delta function to function spaces, i.e., $\int F(p) \Delta(p - p_0) \, dp = F(p_0)$ for any continuous functional F. The terms $\Delta(p + u_i \delta_{x_i} - p')$ and $\Delta(p - u_i \delta_{x_i} - p')$ thus ensure that the transition rate r vanishes unless p' can be reached from p through a birth event (first term) or death event (second term) in species i. The death rate $r_i^-(x_i, p)$ is given by multiplying the per capita death rate $d_i(x_i, p)$ with the density $p_i(x_i)$ of individuals at that trait value,

$$r_i^-(x_i, p) = d_i(x_i, p) p_i(x_i) \,.$$

Similarly, the birth rate $r_i^+(x_i, p)$ at trait value x is given by

$$r_i^+(x_i, p) = [1 - \mu_i(x_i)] b_i(x_i, p) p_i(x_i) + \int \mu_i(x_i') b_i(x_i', p) p_i(x_i') M_i(x_i', x_i) \, dx_i' \,,$$

with the first and second terms corresponding to births without and with mutation, respectively. The master equation above, together with its transition rates, describes so-called generalized replicator dynamics (Dieckmann 1994) and offers a generic formal framework for deriving simplified descriptions of individual-based mutation-selection processes.

Monomorphic Stochastic Model. If the time intervals between successfully invading mutations are long enough for evolution to be mutation-limited, $\mu_i(x_i) \to 0$ for all i and x_i, the evolving populations will remain monomorphic at almost any moment in time (unless and until evolutionary branching occurs). We can then consider trait substitutions resulting from the successful invasion of mutants into monomorphic resident populations that have attained their ecological equilibrium. Denoting trait values and population sizes by x_i and n_i for the residents in species $i = 1, \ldots, N$ and by x_j' and n_j' for a mutant in species j, we can substitute the density $p = (n_1 \delta_{x_1}, \ldots, n_N \delta_{x_N}) + u_j n_j' \delta_{x_j'}$ into the generalized replicator dynamics defined above to obtain a master equation for the probability $P(n, n_j')$ of jointly observing resident population sizes n and mutant population size n_j'.

Assuming that the mutant is rare while the residents are sufficiently abundant to be described deterministically, this master equation is equivalent to the joint dynamics

$$\frac{\mathrm{d}}{\mathrm{d}t} n_i = [b_i(x_i,p) - d_i(x_i,p)] n_i$$

for the resident populations with $i = 1, \ldots, N$ and

$$\frac{\mathrm{d}}{\mathrm{d}t} P(n'_j) = b_j(x'_j,p) P(n'_j - 1) + d_j(x'_j,p) P(n'_j + 1)$$

for the mutant population in species j, where $p = (m_1 \delta_{x_1}, \ldots, m_N \delta_{x_N})$ and $P(m'_j)$ denotes the probability of observing mutant population size m'_j. The rare mutant thus follows a homogeneous and linear birth-death process.

Assuming that the resident community is at its equilibrium, the conditions $b_i(x_i, \bar{p}) = d_i(x_i, \bar{p})$ for all species $i = 1, \ldots, N$ define $\bar{n}_i(x)$ and thus $\bar{p}(x) = (\bar{n}_1(x) \delta_{x_1}, \ldots, \bar{n}_N(x) \delta_{x_N})$, $\bar{b}_j(x'_j, x) = b_j(x'_j, \bar{p}(x))$, $\bar{d}_j(x'_j, x) = d_j(x'_j, \bar{p}(x))$, and $\bar{f}_j(x'_j, x) = \bar{b}_j(x'_j, x) - \bar{d}_j(x'_j, x)$. When the resident population in species j is small enough to be subject to accidental extinction through demographic stochasticity, $s_j(x'_j, x) = (1 - e^{-2 \tilde{f}_j(x'_j, x)})/(1 - e^{-2 \tilde{f}_j(x'_j, x) \bar{n}_j(x)})$ with $\tilde{f}_j(x'_j, x) = \bar{f}_j(x'_j, x))/[\bar{b}_j(x'_j, x) + \bar{d}_j(x'_j, x)]$ approximates the probability of a single mutant individual with trait value x'_i to survive accidental extinction through demographic stochasticity and to go to fixation by replacing the former resident with trait value x_i (e.g., Crow and Kimura 1970). When the resident population in species j is large, $\bar{n}_j(x) \to \infty$, this probability converges to the simpler expression $s_j(x'_j, x) = \max(0, \bar{f}_j(x'_j, x))/\bar{b}_j(x'_j, x)$ known from branching process theory (e.g., Athreya and Ney 1972).

Once mutants have grown beyond the range of low population sizes in which accidental extinction through demographic stochasticity is still very likely, they are generically bound to go to fixation and thus to replace the former resident, provided that their trait value is sufficiently close to that of the resident, $x'_j \approx x_j$ (Geritz et al. 2002). Hence the transition rate $r(x', x)$ for the trait substitution $x \to x'$ is given by multiplying (i) the distribution $\mu_j(x_j) \bar{b}_j(x_j, x) M_j(x'_j, x_j)$ of arrival rates for mutants x'_j among residents x, with (ii) the probability $s_j(x'_j, x)$ of mutant survival given arrival, and with (iii) the probability 1 of mutant fixation given survival,

$$r(x', x) = \sum_{j=1}^{N} \mu_j(x_j) \bar{b}_j(x_j, x) M_j(x'_j, x_j) \bar{n}_j(x) s_j(x'_j, x) \prod_{i=1, i \neq j}^{N} \delta(x'_i - x_i)$$

(Dieckmann 1994; Dieckmann et al. 1995; Dieckmann and Law 1996). Here the product of Dirac delta functions captures the fact that all but the jth component of x remain unchanged, while the summation adds the transition rates for those j^{th} components across all species.

Based on these transition rates, the master equation for the probability $P(x)$ of observing trait value x,

$$\frac{d}{dt}P(x) = \int [r(x,x')P(x') - r(x',x)P(x)]\,dx'\,,$$

then describes the directed evolutionary random walks in trait space resulting from sequences of trait substitutions.

Monomorphic Deterministic Model. If mutational steps $x_i \to x'_i$ are small, the average of many realizations of the evolutionary random walk model described above is closely approximated by

$$\frac{d}{dt}x_i = \int (x'_i - x_i)r(x',x)\,dx'$$

for $i = 1, \ldots, N$ (e.g., van Kampen 1981). After inserting $r(x',x)$ as derived above, this yields

$$\frac{d}{dt}x_i = \mu_i(x_i)\bar{b}_i(x_i,x)\bar{n}_i(x)\int s_i(x'_i,x)(x'_i - x_i)M_i(x'_i,x_i)\,dx'_i\,.$$

By expanding $s_i(x'_i,x) = \max(0, \bar{f}_i(x'_i,x))/\bar{b}_i(x'_i,x)$ around x_i to first order in x'_i, we obtain $s_i(x'_i,x) = \max(0, (x'_i - x_i)g_i(x))/\bar{b}_i(x_i,x)$ with $g_i(x) = \frac{\partial}{\partial x'_i}\bar{f}_i(x'_i,x)\big|_{x_i=x'_i}$; notice here that $\bar{f}_i(x_i,x) = 0$. This means that in the x'_i-integral above only half of the total x'_i-range contributes, while for the other half the integrand is 0. If mutation distributions M_i are symmetric – $M_i(x_i + \Delta x_i, x_i) = M_i(x_i - \Delta x_i, x_i)$ for all i, x_i, and Δx_i – we obtain

$$\frac{d}{dt}x_i = \frac{1}{2}\mu_i(x_i)\bar{n}_i(x)\int (x'_i - x_i)^T(x'_i - x_i)M_i(x'_i,x_i)\,dx'\,g_i(x)\,.$$

The integral is the variance-covariance matrix of the mutation distribution M_i around trait value x_i, denoted by $\sigma_i^2(x_i)$. Hence we recover the canonical equation of adaptive dynamics (Dieckmann 1994; Dieckmann and Law 1996),

$$\frac{d}{dt}x_i = \frac{1}{2}\mu_i(x_i)\bar{n}_i(x)\sigma_i^2(x_i)g_i(x)$$

for $i = 1, \ldots, N$. When mutational steps $x_i \to x'_i$ are not small, higher-order correction terms can be derived: these improve the accuracy of the canonical equation and also cover non-symmetric mutation distributions (Dieckmann 1994; Dieckmann and Law 1996).

Polymorphic Deterministic Model. When mutation probabilities are high, evolution is no longer mutation-limited, so that the two classes of models introduced above – both being derived from the analysis of invasions into essentially monomorphic populations – cannot offer quantitatively accurate approximations of the underlying individual-based birth-death-mutation processes. Provided that population sizes are sufficiently large, it instead becomes appropriate to investigate the average distibution-valued dynamics of many realizations of the birth-death-mutation process,

$$\frac{\mathrm{d}}{\mathrm{d}t}p(x) = \int [p'(x) - p(x)] r(p', p) \, \mathrm{d}p' \,.$$

Inserting the transition rates $r(p', p)$ specified above for the individual-based evolutionary model, we can infer (by collapsing the integrals using the sifting properties of the Dirac delta function and of the generalized delta function)

$$\frac{\mathrm{d}}{\mathrm{d}t}p_i(x) = r_i^+(x_i, p) - r_i^-(x_i, p)$$

for $i = 1, \ldots, N$. Inserting $r_i^+(x_i, p)$ and $r_i^-(x_i, p)$ from above, this gives

$$\frac{\mathrm{d}}{\mathrm{d}t}p_i(x) = [1 - \mu_i(x_i)] b_i(x_i, p) p_i(x_i)$$
$$+ \int \mu_i(x_i') b_i(x_i', p) p_i(x_i') M_i(x_i', x_i) \, \mathrm{d}x_i' - d_i(x_i, p) p_i(x_i) \,.$$

Further analysis is simplified by assuming that the mutation distributions M_i are not only symmetric but also homogeneous – $M_i(x_i' + \Delta x_i, x_i + \Delta x_i) = M_i(x_i', x_i)$ for all i, x_i', x_i, and Δx_i. Expanding $\mu_i(x_i') b_i(x_i', p) p_i(x_i')$ up to second order in x' around x_i,

$$\mu_i(x_i') b_i(x_i', p) p_i(x_i') = \mu_i(x_i) b_i(x_i, p) p_i(x_i) + (x_i' - x_i) \frac{\partial}{\partial x_i} \mu_i(x_i) b_i(x_i, p) p_i(x_i)$$
$$+ \tfrac{1}{2}(x_i' - x_i)^{\mathrm{T}} [\frac{\partial^2}{\partial x_i^2} \mu_i(x_i) b_i(x_i, p) p_i(x_i)](x_i' - x_i),$$

then yields

$$\frac{\mathrm{d}}{\mathrm{d}t}p_i(x) = f_i(x_i, p) p_i(x_i) + \frac{1}{2}\sigma_i^2(x_i) * \frac{\partial^2}{\partial x_i^2} \mu_i(x_i) b_i(x_i, p) p_i(x_i) \,,$$

with $f_i(x_i, p) = b_i(x_i, p) - d_i(x_i, p)$, $\sigma_i^2(x_i) = \int (x_i' - x_i)^{\mathrm{T}} (x_i' - x_i) M_i(x_i', x_i) \, \mathrm{d}x_i'$, and with $*$ denoting the elementwise multiplication of two matrices followed by summation over all resultant matrix elements. This result also provides a good approximation when mutation distributions are heterogeneous, as long as $\sigma_i^2(x_i)$, rather than being strictly independent of x_i, varies very slowly with x_i on the scale given by its elements.

References

1. Arditi, R., and Ginzburg, L.R. (1989). Coupling in predator-prey dynamics: Ratio-dependence. *Journal of Theoretical Biology* 139: 311–326
2. Athreya, K.B., and Ney, P.E. (1972). *Branching Processes*. New York, USA: Springer-Verlag

3. Brännström, Å., Loeuille, N., Loreau, M., and Dieckmann, U. Metabolic scaling, competition, and predation induce repeated adaptive radiation, in preparation
4. Bürger, R. (1998). Mathematical properties of mutation-selection models. *Genetica* 103: 279–298
5. Bush, G.L. (1969). Sympatric host race formation and speciation in frugivorous flies of the genus *Rhagoletis* (Diptera: Tephritidae). *Evolution* 23: 237–251
6. Caldarelli, G., Higgs, P.G., and McKane, A.J. (1998). Modelling coevolution in multispecies communities. *Journal of Theoretical Biology* 193: 345–358
7. Calow, P. (1999). *Encyclopedia of Ecology and Environmental Management.* Oxford, UK: Blackwell Publishing
8. Cheptou, P.O. (2004). Allee effect and self-fertilization in hermaphrodites: Reproductive assurance in demographically stable populations. *Evolution* 58: 2613–2621
9. Christiansen, F.B. (1991). On conditions for evolutionary stability for a continuously varying character. *American Naturalist* 138: 37–50
10. Clements, F.E. (1916). Plant succession: An analysis of the development of vegetation. Publication No. 242. Carnegie Institute Washington, Washington, D.C., USA
11. Crank, J. (1975). *The Mathematics of Diffusion.* Oxford, UK: Clarendon Press
12. Cressman, R. (1990). Evolutionarily stable strategies depending on population-density. *Rocky Mountain Journal of Mathematics* 20: 873–877
13. Crow, J.F., and Kimura, M. (1970). *An Introduction to Population Genetics Theory.* New York, USA: Harper and Row
14. Dercole, F., and Rinaldi, S. (2002). Evolution of cannibalistic traits: Scenarios derived from adaptive dynamics. *Theoretical Population Biology* 62: 365–374
15. Dercole, F., Ferrière, R., and Rinaldi, S. (2002). Ecological bistability and evolutionary reversals under asymmetrical competition. *Evolution* 56: 1081–1090
16. Dieckmann, U. (1994). *Coevolutionary Dynamics of Stochastic Replicator Systems.* Jülich, Germany: Central Library of the Research Center Jülich
17. Dieckmann, U., and Doebeli, M. (1999). On the origin of species by sympatric speciation. *Nature* 400: 354–357
18. Dieckmann, U., and Ferrière, R. (2004). Adaptive dynamics and evolving biodiversity. In: Ferrière, R., Dieckmann, U., and Couvet, D., eds. *Evolutionary Conservation Biology.* Cambridge, UK: Cambridge University Press, pp. 188–224
19. Dieckmann, U., and Law, R. (1996). The dynamical theory of coevolution: A derivation from stochastic ecological processes. *Journal of Mathematical Biology* 34: 579–612
20. Dieckmann, U., Doebeli, M., Metz, J.A.J., and Tautz, D. eds. (2004). *Adaptive Speciation.* Cambridge, UK: Cambridge University Press
21. Dieckmann, U., Marrow, P., and Law, R. (1995). Evolutionary cycling in predator-prey interactions: Population dynamics and the Red Queen. *Journal of Theoretical Biology* 176: 91–102
22. Diehl, S., and Feissel, M. (2000). Effects of enrichment on three-level food chains with omnivory. *American Naturalist* 155: 200–218
23. Doebeli, M., and Dieckmann, U. (2000). Evolutionary branching and sympatric speciation caused by different types of ecological interactions. *American Naturalist* 156: S77–S101

24. Doebeli, M., and Dieckmann, U. (2003). Speciation along environmental gradients. *Nature* 421: 259–264
25. Doebeli, M., and Dieckmann, U. (2005). Adaptive dynamics as a mathematical tool for studying the ecology of speciation processes. *Journal of Evolutionary Biology* 18: 1194–1200
26. Doebeli, M., Dieckmann, U., Metz, J.A.J., and Tautz, D. (2005). What we have also learned: Adaptive speciation is theoretically plausible. *Evolution* 59: 691–695
27. Drake, J.A. (1990). Communities as assembled structures: Do rules govern pattern? *Trends in Ecology and Evolution* 5: 159–164
28. Drossel, B., and McKane, A.J. (2000). Competitive speciation in quantitative genetic models. *Journal of Theoretical Biology* 204: 467–478
29. Drossel, B., Higgs, P.G., and McKane, A.J. (2001). The influence of predator-prey population dynamics on the long-term evolution of food web structure. *Journal of Theoretical Biology* 208: 91–107
30. Eliot, C. Method and metaphysics in Clements's and Gleason's ecological explanations. *Studies in History and Philosophy of Science Part C: Studies in History and Philosophy of Biological and Biomedical Sciences*, in press (available online at http://people.hofstra.edu/faculty/Christopher_H_Eliot/EliotClements.pdf)
31. Elton, C.S. (1958). *Ecology of invasions by animals and plants*. London, UK: Chapman and Hall
32. Ernande, B., Dieckmann, U., and Heino, M. (2002). Fisheries-induced changes in age and size at maturation and understanding the potential for selection-induced stock collapse. *ICES CM* 2002/Y:06
33. Eshel, I. (1983). Evolutionary and continuous stability. *Journal of Theoretical Biology* 103: 99–111
34. Felsenstein, J. (1981). Skepticism towards Santa Rosalia, or Why are there so few kinds of animals? *Evolution* 35: 124–238
35. Ferrière, R. (2000). Adaptive responses to environmental threats: Evolutionary suicide, insurance, and rescue. International Institute for Applied Systems Analysis, Laxenburg, Austria: *Options* Spring 2000, pp. 12–16
36. Ferrière, R., Bronstein, J.L., Rinaldi, S., Gauduchon, M., and Law, R. (2002). Cheating and the evolutionary stability of mutualism. *Proceedings of Royal Society of London Series B* 269: 773–780
37. Fisher, R.A. (1930). *The Genetical Theory of Natural Selection*. Oxford, UK: Clarendon Press
38. Geritz, S.A.H., and Kisdi, É. (2000). Adaptive dynamics in diploid sexual populations and the evolution of reproductive isolation. *Proceedings of the Royal Society of London B* 267: 1671–1678
39. Geritz, S.A.H., Gyllenberg, M., Jacobs, F.J.A., and Parvinen, K. (2002) Invasion dynamics and attractor inheritance. *Journal of Mathematical Biology* 44: 548–560
40. Geritz, S.A.H., Kisdi, É., Meszéna, G., and Metz, J.A.J. (1998). Evolutionary singular strategies and the adaptive growth and branching of the evolutionary tree. *Evolutionary Ecology* 12: 35–57
41. Geritz, S.A.H., Kisdi, É., Meszéna, G., and Metz, J.A.J. (2004). Adaptive dynamics of speciation: Ecological underpinnings. In: Dieckmann, U., Doebeli, M., Metz, J.A.J., and Tautz, D. eds. *Adaptive Speciation*. Cambridge, UK: Cambridge University Press, pp. 54–75

42. Geritz, S.A.H., Metz, J.A.J., Kisdi, É., and Meszéna, G. (1997). Dynamics of adaptation and evolutionary branching. *Physical Review Letters* 78: 2024–2027
43. Gillespie, D.T. (1976). A general method for numerically simulating the stochastic time evolution of coupled chemical reactions. *Journal of Computational Physics* 22: 403–434
44. Gleason, H.A. (1926). The individualistic concept of the plant association. *Bulletin of the Torrey Botanical Club* 53: 7–26
45. Gyllenberg, M., and Parvinen, K. (2001). Necessary and sufficient conditions for evolutionary suicide. *Bulletin of Mathematical Biology* 63: 981–993
46. Gyllenberg, M., Parvinen, K., and Dieckmann, U. (2002). Evolutionary suicide and evolution of dispersal in structured metapopulations. *Journal of Mathematical Biology* 45: 79–105
47. Haldane, J.B.S. (1932). *The Causes of Evolution*. London, UK: Harper
48. Hardin, G. (1968). The tragedy of the commons. *Science* 162: 1243–1248
49. HilleRisLambers, R., and Dieckmann, U. (2003). Competition and predation in simple food webs: Intermediately strong trade-offs maximize coexistence. *Proceedings of the Royal Society of London B* 270: 2591–2598
50. HilleRisLambers, R., and Dieckmann, U. Evolving Omnivory: Restrictions on simple food webs imposed by the interplay between ecology and evolution, submitted
51. Holt, R.D., and Polis, G.A. (1997). A theoretical framework for intraguild predation. *American Naturalist* 149: 745–764
52. Hubbell, S.P. (2001). *The Unified Neutral Theory of Biodiversity and Biogeography*. Princeton, USA: Princeton University Press
53. Ito, H., and Ikegami, T. (2003). Evolutionary dynamics of a food web with recursive branching and extinction. In: Standish, R.K., Bedau, M.A., and Abbass, H.A. eds. *Artificial Life VIII*, Cambridge, USA: MIT Press, pp. 207–215
54. Ito, H.C., and Ikegami, T. (2006). Food web formation with recursive evolutionary branching. *Journal of Theoretical Biology* 238: 1–10
55. Johnson, P.A., Hoppensteadt, F.C., Smith, J.J., and Bush, G.L. (1996). Conditions for sympatric speciation: A diploid model incorporating habitat fidelity and non-habitat assortative mating. *Evolutionary Ecology* 10: 187–205
56. Kimura, M. (1965). A stochastic model concerning maintenance of genetic variability in quantitative characters. *Proceedings of the National Academy of Sciences of the USA* 54: 731–735
57. Kirkpatrick, M. (1996). Genes and adaptation: A pocket guide to theory. In: Rose, M.R., and Lauder, G.V. eds. *Adaptation*. San Diego, USA: Academic Press, pp. 125–128
58. Kisdi, É., and Meszéna, G. (1993). Density-dependent life-history evolution in fluctuating environments. In: Yoshimura, J., and Clark, C. eds. *Adaptation in a Stochastic Environment*, Lecture Notes in Biomathematics 98. Berlin, Germany: Springer, pp. 26–62
59. Kokko, H., and Brooks, R. (2003). Sexy to die for? Sexual selection and the risk of extinction. *Annales Zoologici Fennici* 40: 207–219
60. Law, R. (1999). Theoretical aspects of community assembly. In: McGlade, J. ed. *Advanced ecological theory: Principles and applications*. Oxford, UK: Blackwell Science, pp. 143–171
61. Law, R., Marrow, P., and Dieckmann, U. (1997). On evolution under asymmetric competition. *Evolutionary Ecology* 11: 485–501

62. Le Galliard, J.F., Ferrière, R., and Dieckmann, U. (2003). The adaptive dynamics of altruism in spatially heterogeneous populations. *Evolution* 57: 1–17
63. Levins, R. (1962). Theory of fitness in a heterogeneous environment. I. The fitness set and adaptive function. *American Naturalist* 96: 361–373
64. Levins, R. (1968). *Evolution in Changing Environments*. Princeton, USA: Princeton University Press
65. Loeuille, N., and Loreau, M. (2005). Evolutionary emergence of size-structured food webs. *Proceedings of the National Academy of Sciences of the USA* 102: 5761–5766
66. MacArthur, R., and Levins, R. (1967). The limiting similarity, convergence, and divergence of coexisting species. *American Naturalist* 101: 377–385
67. Matsuda, H. (1985). Evolutionarily stable strategies for predator switching. *Journal of Theoretical Biology* 115: 351–366
68. Matsuda, H., and Abrams, P.A. (1994a). Runaway evolution to self-extinction under asymmetrical competition. *Evolution* 48: 1764–1772
69. Matsuda, H., and Abrams, P.A. (1994b). Timid consumers – Self-extinction due to adaptive change in foraging and anti-predator effort. *Theoretical Population Biology* 45: 76–91
70. May, R.M. (1973). *Stability and Complexity in Model Ecosystems*. Princeton, USA: Princeton University Press
71. Maynard Smith, J. (1966). Sympatric speciation. *American Naturalist* 100: 637–650
72. Maynard Smith, J. (1982). *Evolution and the Theory of Games*. Cambridge, UK: Cambridge University Press
73. Maynard Smith, J., and Price, G.R. (1973). Logic of animal conflict. *Nature* 246: 15–18
74. Mayr, E. (1963). *Animal Species and Evolution*. Cambridge, USA: Harvard University Press
75. Mayr, E. (1982). *The Growth of Biological Thought: Diversity, Evolution, and Inheritance*. Cambridge, USA: The Belknap Press of Harvard University Press
76. McCann, K.S. (2000). The diversity-stability debate. *Nature* 405: 228–233
77. Meszéna, G., Kisdi, É., Dieckmann, U., Geritz, S.A.H., and Metz, J.A.J. (2001). Evolutionary optimisation models and matrix games in the unified perspective of adaptive dynamics. *Selection* 2: 193–210
78. Metz, J.A.J., Geritz, S.A.H., Meszéna, G., Jacobs, F.J.A., and van Heerwaarden, J.S. (1996). Adaptive dynamics, A geometrical study of the consequences of nearly faithful reproduction. In: van Strien, S.J., Verduyn Lunel, S.M. eds. *Stochastic and Spatial Structures of Dynamical Systems*. Amsterdam, The Netherlands: North-Holland, pp. 183–231
79. Metz, J.A.J., Nisbet, R.M., and Geritz, S.A.H. (1992). How should we define fitness for general ecological scenarios? *Trends in Ecology and Evolution* 7: 198–202
80. Meyer, A. (1993). Phylogenetic relationships and evolutionary processes in east African cichlid fishes. *Trends in Ecology and Evolution* 8: 279–284
81. Mylius, S.D., Klumpers, K., de Roos, A.M., and Persson, L. (2001). Impact of intraguild predation and stage structure on simple communities along a productivity gradient. *American Naturalist* 158: 259–276
82. Odling-Smee, F.J., Laland, K.N., and Feldman, M.W. (2003). *Niche Construction: The Neglected Process in Evolution*. Princeton, USA: Princeton University Press

83. Oksanen, L., and Oksanen, T. (2000). The logic and realism of the hypothesis of exploitation ecosystems. *American Naturalist* 155: 703–723
84. Oksanen, L., Fretwell, S., Arruda, J., and Niemela, P. (1981). Exploitation ecosystems in gradients of primary productivity. *American Naturalist* 118: 240–261
85. Parvinen, K. (2006). Evolutionary suicide. *Acta Biotheoretica* 53: 241–264
86. Peters, R.H. (1983). *The Ecological Implications of Body Size*. Cambridge, UK: Cambridge University Press
87. Post, W.M., and Pimm, S.L. (1983). Community assembly and food web stability. *Mathematical Biosciences* 64: 169–192
88. Press, W.H., Teukolsky, S.A., Vetterling, W.T., and Flannerty, B.P. (1992). *Numerical Recipes in C: The Art of Scientific Computing*, 2nd edition. Cambridge, USA: Cambridge University Press
89. Rankin, D.J., and López-Sepulcre, A. (2005). Can adaptation lead to extinction? *Oikos* 111: 616–619
90. Rosenzweig, M.L. (1978). Competitive speciation. *Biological Journal of the Linnean Society* 10: 275–289
91. Roughgarden, J. (1974) Species packing and the competition function with illustrations from coral reef fish. *Theoretical Population Biology* 5: 163–186
92. Roughgarden, J. (1979). *Theory of Population Genetics and Evolutionary Ecology: An Intro-duction*. New York, USA: Macmillan
93. Roughgarden, J. (1983). The theory of coevolution. In: Futuyma, D.J., and Slatkin, M. eds. *Coevolution*. Sunderland, USA: Sinauer Associates, pp. 33–64
94. Rummel, J.D., and Roughgarden, J. (1985). A theory of faunal build-up for competition communities. *Evolution* 39:1009–1033
95. Schliewen, U.K., Tautz, D., and Pääbo, S. (1994) Sympatric speciation suggested by monophyly of crater lake cichlids. *Nature* 368: 629–632
96. Schluter, D. (2000). *The Ecology of Adaptive Radiation*. Oxford, UK: Oxford University Press
97. Taper, M.L., and Case, T.J. (1992). Coevolution among competitors. In: Futuyma, D., and Antonivics, J. eds. *Oxford Surveys in Evolutionary Biology, Volume 8*. Oxford, UK: Oxford University Press. pp. 63–111
98. Tilman, D. (1982). *Resource Competition and Community Structure*. Princeton, USA: Princeton University Press
99. Udovic, D. (1980). Frequency-dependent selection, disruptive selection, and the evolution of reproductive isolation. *American Naturalist* 116: 621–641
100. van Kampen, N.G. (1992). *Stochastic Processes in Physics and Chemistry*. Amsterdam, The Netherlands: North-Holland
101. van Tienderen, P.H., and de Jong, G. (1986). Sex-ratio under the haystack model – Polymorphism may occur. *Journal of Theoretical Biology* 122: 69–81
102. Vandermeer, J.H. (1970). The community matrix and the number of species in a community. *American Naturalist* 104: 73–83
103. Webb, C.T. (2003). A complete classification of Darwinian extinction in ecological interactions. *American Naturalist* 161: 181–205
104. Wright, S. (1932). The roles of mutation, inbreeding, crossbreeding and selection in evolution. *Proceedings of the 6th International Congress of Genetics* 1: 356–366
105. Wright, S. (1967). Surfaces of selective value. *Proceedings of the National Academy of Sciences of the USA* 102: 81–84

Index

accidental extinction 170
action 27
adaptation 146, 167
adaptive dynamics theory 147
adaptive radiations 165, 167
adaptive speciation 153
advantage of rarity 152
age-structured model 66
    age-structured model with time delay 68
allee effects 56, 155
allopatric speciation 153
altruism 21
assembly models 146
assessment 25, 26
assortative mating 152
asymmetric competition 165
asynchronous entry 36, 37
attack coefficients 157, 158
attractor 133
    chaotic 125
    2-cycle 128
    lattice 137
attractors 125, 127, 135, 138
    cycle 138
    multiple 127, 133
autotrophic 160

basal resource 156
basic reproduction ratio 15
basin 133
    hopping 133
    of attraction 133

bifurcations 127
biofilms 94
birth-death process 147, 170
birth-death-mutation process 149, 168
body size 160
Brownian motion 10

cannibalism 6
canonical equation 157, 171
canonical equation of adaptive dynamics 149, 171
catastrophic bifurcation 155
cell 93
    adherent 96
    planktonic 96
    plasmid-bearing 94
    plasmid-free 94
chaos 127, 135, 136
    control of 127
Chapman-Kolmogorov 9
characteristic equation 17
chemostat 97
Clements 145
Clements-Gleason continuum 146
Clements-Gleason debate 145
coevolutionary attractor 158
coexisting attractors 148
cohort cycles 6
community complexity or diversity 146
community stability 146
community stability or productivity 146

community structures 145, 168
competition 127
competition kernel 161
competition parameters 162
competitive speciation 153
competitive system 75
  autonomous competitive system 75
  periodic competitive system 76
concatenation 8
conjugation 93
constant fitness landscapes 146
consumer species 156
continuous transition to
  extinction 155
convergence stability 148, 152
conversion efficiencies 157
cooperation 21
cooperative matrix 54
correction terms 171
costly punishment 25, 26
cycle 127, 147
  attenuant 134
  periodic 127
2-cycle 128

Darwinian extinction 154
defections 23
demographic stochasticity 170
density-dependent selection 153
destabilization 127
diffusion 164
dimorphism 152
Dirac's delta function 147, 150, 165, 168
directional evolution 165
directional selection 167
discontinuous transition to
  extinction 155
dispersal delay 62
disruptive selection 152
dynamics 32, 140
  chaotic 140
  lattice 140

e-commerce 22
ecological communities 145
ecological niches 166
ecological opportunities 167
ecological speciation 153

ecological stability 156
energy parameters 162
environmental condition 5
equilibria 127
errors 21, 34, 37
evolutionarily singular trait
  values 148
evolutionarily stable strategies
  (ESS) 29
evolutionary branching 162, 165, 167
evolutionary branching point
  152, 167
evolutionary cycles 159
evolutionary dynamics 147
evolutionary game dynamics 22
evolutionary game theory 21, 153
evolutionary random walk 149, 171
evolutionary random walks 149, 167
evolutionary stability 148, 152, 156
evolutionary suicide 154, 155
experiments 43
exponent 133
  Lyapunov 133, 140
extinction 154, 167

feedback 5
feeding preferences 157
Fisher 154
fitness functions 147
fitness landscapes 146
fitness minimum 152, 153
fixation 170
fixed points 32
fluctuating environments 148
flyby 138
folk theorem 22
food chain 157, 158
food web 156, 158
foraging kernel 161
foraging parameters 162
frequency- and density-dependent
  selection 147
frequency-dependent selection
  152–154
full score 42
functional response 157, 164
fundamental theorem of natural
  selection 154

games
    evolutionary 24
    experimental 43
    spatial 41
generalist advantage 158
generalized delta function 169
generalized replicator dynamics 169
geographical isolation 153
Gillespie 149
Gleason 145
gradient-ascent model 149, 167

habitat 140
Haldane 154
Hardin 154
heterotrophic 160

immigrants 146
immigration 146, 167
indirect observation model 29
individual 5
input 8
inshore-offshore fishery 69
interacting particle systems 81
interaction 146, 167
interference competition 160, 162
intraguild predation 156
invasion fitness 148, 152

kernel 8
Kronecker delta symbols 169

lag metric 138
lattice 127
    effects 135
lattice models 81
leading eight 21, 28
learning process 24
Levins 154
Lotka–Volterra 166
Lotka–Volterra dynamics 157, 160

manifold 128
    stable 128, 129
    unstable 128, 129
mass-action 12
master equation 168, 170
matrix games 153
maturation 5

mean residence time 98
mean-field approximation 83
memory 25
metabolic scaling 162
metapopulation 51
    metapopulation model 57
    source-sink metapopulation 52
minimal process method 149
mixture 39
model
    "Poisson/binomial" LPA 139
    deterministic 125
    lattice 136
    LPA 128
    probabilistic 138
    stochastic 133, 135
monomorphic 147, 169
monomorphic deterministic
        model 171
monomorphic stochastic
        model 169
moral hazard 22
moral judgement 26
moral systems 22
morals 27, 46
multi-locus models 155
mutant Hessian 148
mutation 24, 147
mutual intra-guild predation 157

neutral theory 145
niche construction 145
noise 126, 131, 135, 137
non-equilibrium ecological
        dynamics 148
nonlinearity 125
normal form 152, 155
number of rounds 36, 38

oligomorphic 147, 161
omnivory 156
optimizing selection 146, 152, 154

pair approximation 84
pairwise invasibility plots 148
patterns 136
    deterministic 136
periodic system 58
permanence 59

182 Index

persistence 51
  partial permanence 65
  partially permanent 63
  partially persistent 63
perturbations 126
  stochastic 126, 129, 134
phenotypic clusters 165
physiological population
    structure 148
plasmids 93
polymorphic deterministic model 171
polymorphic stochastic model 168
polymorphisms 40
population 5
predator-prey system 71
  time-dependent predator-prey
    system 73
premating isolation 152
public good game 45
punishment 21, 45

quantitative traits 147
quasi-periodic 127

R^* rule 156
random communities 145
random drift 32
ratio-dependent functional
    response 165
reaction-diffusion model 149
reciprocity
  direct 21
  indirect 21, 24
recombination 152
recurrent evolutionary
    branching 163
renewal equation 13
replicator dynamics 30, 36
replicator equation 32, 153
reproduction 5
reproductive isolation 152
reputation 22
resident Hessian 148
resonance 127, 128, 134
Ricker map 133
  stochastic 133
route-to-chaos 127, 135
runaway evolution to
    self-extinction 155

saddle 128, 135, 140
  flyby 128, 129, 134
score 24
scoring 46
second order social dilemma 26
segregative loss 93
selection pressures 147
selection-driven deterioration 155
selection-driven extinction 147,
    154, 155
semichemostat dynamics 157
semigroup 13
sequential evolutionary branching 161
sexual inheritance 155
sexual populations 152
sexual reproduction process 82
sifting property 168, 169
simplex 32
simulations 40, 41, 136
  stochastic 136
single-species system 52
skeleton 139
  deterministic 139
social dilemma 25
social network 21, 36, 38
social norm 26
spatially distibuted populations 41
specialist advantage 158
speciation 147, 152
species pool 146
stability 54, 127
  globally stable 56
stability modulus 98
standing 23, 25, 46
steady state 15
stirring 82
stochasticity 125, 132, 137, 142
  demographic 132, 138, 141
strategy 24, 26
strong reciprocation 26
survival 5
sympatric speciation 153

Tilman 156
time scales 146
trade-off 157, 158
tragedy of the commons 154
trait substitution 149, 152, 154
transcritical bifurcation 155

transients   133, 135, 136, 140
transition rate   149, 169, 170, 172
Tribolium castaneum   128, 134
trigger strategies   22
triplet decoupling approximation   86
trophic efficiency   161
trophic interactions   158, 163

utilization spectrum   163

variance-covariance matrix   171

weak patchy environment   58
   food-poor   60
   food-rich   60
Wright   154

DATE DUE

DUE DATE SUBJECT TO CHANGE
IF A RECALL IS REQUESTED